Novel: Marvel to Excel

LIFE TIDES

(A Novel on the Plight of Farmers in the Parody)

Dr. RAJAMOULY KATTA

Creative Crystals
PUBLISHING

Dallas, TX, USA

DEDICATION:
TO FARMERS

ACKNOWLEDGMENTS

I wholeheartedly thank Prof. Sreelatha Tangirala, Department of English, for her excellent foreword to the novel, **Life Tides**. I at the same time thank the publishers, for the neat publication of the novel.

Dr. Rajamouly Katta

PREFACE

Poets, playwrights and other crative writers in the past presented the kaleidoscopic picture of the past in their works as they had creative commitment to respond to the virtues and react to the vices of the people in the society, they lived in. In the past, there were demons and humans, and the humans became victims to the demons. Finally, the humans fought against the demons with the help of Divines and reaped victory in Dwapara Yuga and Treta Yuga.

In the current stage, there are clashes between man-guised demons or cunning humans and innocent humans. There are battles or conflicts between betweem them. Among them, innocents became victims to the fellow cunning humans. The situation prevails in the present scenario in the Kali Yuga.

In the Ramayana, Sri Rama, facing ordeals all through his life, wins the battle against the demon, Ravana with ten heads. In the Mahabharatha, the Pandavas, facing untoward incidents and defeating the hundred Kauravas in Kurukshetra battlefield, came to rule with the help of Lord Krishna. In the present era, evils and vices outnumber virtues and values. My novel, entitled "Life Tides" presents the kaleidoscopic details of the lives of innocent workers, especially farmers. This novel presents the pathetic lives of farmers to its readers.

We find demons delineated in classics and epics in the past exist in the human form in the present society. There are more demons in actions everywhere than humans existing here and there and a few divines like Sri Rama and Yudhistira who are the rarest of the rare in the new millennium.

Classics, epics, and plays written once have relevance in the present context. For example, Shakespeare's plays reflect their relevance in the present scenario, as their characters are at once individual and universal. We find the Hamlets and Claudiuses, the King Lears and the Cordelias, and so on in the present social milieu in the human form. So is the case with the Ravanas and the Kauravas to exit in the human form.

The present witnesses its relevance of the past in respect of human relationships, ideal governance and so on. It feels sorry for the devilish evils in the race of man.

In the literary quest, I study the classics of great poets like Kalidasa to survey their relevance in the present context. I found the relevance of the epic, ***Uttara Ramayanam*** of Valmiki in the present scenario. I wrote a parody of the epic entitled ***Ebb and Flow*** (Yking Books, 2019). On the same lines, I studied ***Abhignana Shakuntam*** of Kalidasa and attempted to write its parody entitled ***Life Tides***.

Dr. Rajamouly Katta

FOREWORD

***"LIFE TIDES* as a Novel Proves to Be a Value Addition to the Lovers of Literature"**
By Prof. Sreelatha Tangirala

* * *

Life Tides a debut novel by Rajamouly Katta, is a touching account of the farmers who are the reliable pillars of an agriculture-based nation like India. The sincere attempt made is truly praiseworthy, which is filled with indigenous flavor. The setting, characters, plot, human relations, background and influences on the novel are truly native; spilling out with local psyche. The style is obviously lucid, simple and realistic in tone and mood.

The opening scene lures with the color and smell of agriculture, the language of farmers and the mood of farming. The words of one of the important characters of the novel, Bharat, show the author's fondness and reverence to farmers and the noble art of farming.

"I'm a farmer...I'm a farmer's son...I'm a farmer's grandson. I'm for the farmer... for the cause of the farmer. When the farmer is happy in a nation, the nation feels pride and dignity. The famer should be happy in a nation like India. The farmers should never shed tears. They shouldn't commit suicides. All should be happy. All should flourish in our nation." (P-126)

Rural mind sets, country politics, cunningness, innocence, and ethics of life are ably dealt with by the author. Suryasen's lust for power with immoral sly political strategies is well introduced in the beginning of the novel which has its phenomenal consequences in the later part of the novel. With such sustained interest, the author tries to surprise the readers with marvelous descriptions and sensible discourses with nature; which obviously reflects the inner voice of the author.

The simple language and realistic nature of the characters like Suryasen, Indrasen, Parameshwar, Vishesh, Malli and Vijay have created required impact on the minds of the readers furnishing a true local feel. Family politics and relations are naturally depicted. The author's eye for the welfare of the villages, farming and farmers, orphans, ideal societal set-ups, model education system is expressed through the plot and characterization of Parameshwar, Vishesh, Shashima, Prashanth and Dakshinya Deekshit. The noble thoughts of the author are heard from the words of Bharat, the Prime Minister of the nation. The idea 'welfare of farmers is the welfare of the Nation' is felt across the novel, till the end.

"Since our nation is an agricultural country, hereafter, we all are farmers having equal share in the land available in India. All must farm the land. Work is worship. Work in the sign of farming is real worship in a nation like India." (P-126)

The plot upholds and elevates the respectful position of women and the displays the writer's attitude towards women and their value in the society. The character 'Malli' serves as a clear reference to the author's words mentioned below:

"Woman is on par man in all respects. She should be happy enjoying freedom. Her race shouldn't become victims to rapes...nor to gang-rapes... not to any humiliation." (P-127)

The writer sincerely attempts to prove Nature as the best teacher and healer. The healing influence of Nature on the minds of the orphans in the orphanages is marvelously depicted. The narration was so poignant and natural signifying the author's love for Nature. The discourses he presented with Nature through Shashima are so alluring.

The author's vision for a better and safe environment on this green planet can be observed as follows:

"If all lands in other countries turn into the sites of houses also, there won't be lands for farming. All roadside lands are with boards of Real Estate Business with colorful banners and attractive ads now. Once there were fields to offer us grains for our food. When no farmers are there, there is no food for all to eat," (P-116)

The technique of using songs in between is another fascinating factor that adds to the beauty of the plot, narration and the indigenous flavor in a very unique manner. It reveals the author's skill in giving a gripping feel to the novel. Another interesting feature worth mentioning is the author's keenness in naming his important characters like Shashima (meaning the glow to Moon), Prashant (meaning peace), Dakshinya Deekshit (Kind and Generous), Karshak Kshetra (Land of Farmers), Vishesh (Unique), Malli (a beautiful-scented flower) and finally Bharat to suit the plot, situations and the personalities.

There is a clear impact of the Indian Puranas and Ithihasas on the thinking of the author. It can be observed in the episode of Shashima and Prashanth. The incidents relating to Shashima and Prashanth resemble exactly the incidents that occur between Shakuntala and Dushyanta. Their parting, Prashant's amnesia, Shashima's stay in

Harsha Deekshit's orphanage and giving birth to a baby boy, naming him Bharat, etc, totally match the happenings of Kalidasa's 'Abhignana Sakuntalam.' Ultimately, when all the things that have fallen apart are set in proper line with the series of the events of the plot, Bharat becomes the commendable and admirable Prime Minister of the nation to the pleasure of all his people far and wide.

The following words reveal the commitment of Bharat as a farmer turned leader, a visionary Prime Minister, the pride of the nation.

"I'm for the people...for their welfare...They'll live peacefully, comfortably, and happily in my governance...I'm 'leader' to mean my people, the members of my family." (P-126)

Thus, this novel proves to be a value addition to the lovers of literature; especially to those who adore indigenous sentiments and feelings. It's a timely and right attempt made by the author to promote awareness among today's youth regarding the selfless and unprofitable but noble efforts of farmers. For a generation that's busily racing towards technology, this book will be an excellent eye opener. It's a thought-provoking gift from the author to the young generation in choosing their careers. They can proudly become educated farmers and improve farming by using modern tools and technology effectively. Bharat, in the novel is the right example to prove that an educated farmer turned leader is the best choice for a nation which respects agriculture as its main occupation but strides with equal pace in the field of technology too.

1

LIFE TIDES

(NOVEL)

The sun rose and was shooting its bright rays in all directions alike. Meanwhile, the clouds spread and were trying their best to hide the invincible rays. The sun made efforts to shatter the glooms, shrouded by clouds. At that moment, the sun was sorry because the clouds were shrouding its bright rays. It was just cloudy weather and so it did not seem to rain.

The people of Indrapur anxiously looked at the sky for the dark-blue clouds to rain as they were farmers to welcome rain to their lives. The clouds appeared to have no wombs of rain at that time.

The farmers of the village, Indrapur had the habit of looking at the sky for they depend on the dark blue clouds that bring rains to turn their lands fertile for good yielding. They, therefore, wished to have rain early in the monsoon.

The sun was trying at its best to come out of the clouds, but it was not possible to do so. It was failing to show its bright rays and falling victim to the clouds at that time.

The farmers felt sorry for their dear sun's helpless position, and were expressing their feelings:

"The clouds are hiding the sun that offers light and warmth to the earth for the mirth of people in its selfless service."

"The invincible rays are badly fated at certain times. The sun has to admit its defeat in the ways."

"The ever-glowing sun bows its head to the stride and ride of clouds."

"The sun bows its head to solar eclipse. The moon fails to scatter its beams during the time of lunar eclipse All greats like the sun and the moon are also in troubles like that."

"Yes, the complete solar eclipse swallows the sun to turn the world into gloom. It is a pity that the sun is swallowed in part or full!"

"The complete lunar eclipse swallows the moon in part or full...We feel pity for the moon in its helpless condition!"

"The sun and the moon struggle to shine for a time. All of us feel sorry for that."

"Yes, the sun burns all the time, and sees the ocean water evaporate to form clouds and the clouds bring us rain in their wombs for our lives."

"Exactly...the sun is responsible for rains. The lives of farmers are indeed a gamble in rain."

"The sun with all great powers sometimes feels that it has misfortunes like us living in the village. We're after all farmers, becoming victims of famines and droughts when there is no rain."

"We, the innocent farmers, are after all puppets in the hands of fate and their fate is changed by the rainfall every year."

"Yes, our lives are rain dependent."

"We at the same time face various atrocities and cruelties of fellow beings."

" Man-to-man relationships are missing nowadays."

"The innocent people are falling victims to the cunning. So is the case with the small farmers as they are exploited by big farmers."

The farmers, owning the small tracks of lands in the village, Indrapur had a problem with a feudal with a hundred acres of land. He was not content with the vast tracks of land he had. He grabbed the lands of small farmers by telling cock and bull stories. He played with the lives of small farmers. He was none other than Suryasen. He was like a despotic king to the people of the village.

Small farmers were cultivating their small tracks of land and were living on agriculture only. Their incomes from agriculture were megre and within the limits.

The farmers lived in the lap of nature, depending on what it offered as free gifts to them. They enjoyed the beauty of nature for the gaieties in their lives.

The small farmers had artisan wells to supply water to their fields. They had healthy oxen to plough their fields. They worked from dawn to dusk and were happy. They were happy with what they had and earned in agriculture. They did not know luxuries as they did not crave them. That was their lifestyle.

Suryasen was such a feudal. He always planned to grab the small tracks of lands from the small farmers. He had the evil goal in the heart of his heart. He was in pursuit of his goals. As a result, they were falling victim to him.

The small farmers were innocent, and they were to respect the feudal for owning a hundred acres of land. For them, he was a big farmer, 'zamindar' excelling them in all respects. They sought his help in forced circumstances. To meet their necessities like children's marriages and education, they borrowed loans from him, mortgaging their small tracks of lands. Almost all the small farmers approached him for some need or the other.

The small farmers did not know the conditions and compulsions mentioned in the agreements the small farmers signed for him. They innocently thought that he was their big farmer and well-wisher.

Suryasen was always in all his efforts to be successful in his selfish mission by means of tactics and knacks.

The small farmers were not able to clear the loans when they were in financial crisis. They agreed to follow Suryasen's suggestions in that crisis.

In the passage of time, Suryasen owned the lands of the small farmers as per the agreements whenever they took loans. They did not know exactly what was happening to them as they were innocent. He gave them assurances:

"You can work in the fields as you've done all the years. I look after your family. You'll not face any problem in your lives."

"I'm here for you. I'm at your disposal."

"This is Suryasen. I'm your man."

"Don't forget me, your well-wisher."

Like that, Suyasen illegally owned the lands of the small farmers. The farmers did not know what exactly had happened to them. They were not able to foresee that their lives would be affected adversely.

They did not expect adversity in their lives at any cost.

Suryasen was a big farmer with no heart. He treated all others as small farmers. Later he grew to be despotic in true spirit due to owning excess lands. He started to dominate the small farmers. treating them as slaves.

Small farmers became laborers to work in their own fields. They innocently obeyed him, as he was a dictator like the despotic king to treat his people as slaves in his rule. No man went against his wishes and orders.

Suryasen treated the small farmers as coolies, laborers, and slaves rather than as fellow farmers though they were farmers working in his fields.

The small farmers sometimes did not like Suryasen's attitude and so they expressed their feelings of protest. Then he manhandled them, using the language of abuses,

"Rogues, you can approach anyone and tell him what you want to tell...None will come to your rescue... Nothing happens to me. Understand... Shut up and get lost soon."

"None can defeat me. I know only success...never a failure."

The small farmers had no other way except to take the abuses easy. They stopped to protest him later at any cost.

When some others slightly protested Suryasen, he beat them severely, saying,

"Fools, you don't know how to live and earn livelihood. You're like sheep to graze in my lands."

"When I'm meeting your dire needs and primary necessities, you don't realize my sincere help to you. You've no mind and so you do not know my greatness."

"Can you give me whatever I want? Never...you after all depend on me in all respects. You are not self sufficient. It's never possible for you to be self sufficient. The noted fact is that you depend on me, but I don't depend on you."

Suryasen used to throw a volley of abuses on the small farmers. His son, Indrasen sometimes beat them indiscriminately.

When the small farmers like the common people did not know what to do, they were living like slaves. Their lives were going on like that.

The nation that strived and craved freedom to result in liberty, equality and fraternity was witnessing violence against its ideals. It wished that principles and morals should exist and persist everywhere.

In the nation, feudal farmers like Suryasen lived and planned for reaching their selfish goals of owning powers and minting treasures. That was the situation prevailed in the peace-loving nation.

There was a piece of news daily in every daily newspaper about such feudal farmers everywhere, especially Suryasen. All the people of virtues felt sorry for the untoward happenings.

Incidents of violence were occurring in almost all places in the nation although it was in quest of peace.

There were people of virtues like Parameshwar, a model farmer and close relative of Suryasen at the neighboring hamlet, Karshak Kshetra. He saw the news item in all newspapers that his relative Suryasen was treating the small farmers as his slaves and felt extremely sorry.

Parameshwar, a farmer with more than a hundred acres of land, treated his workers as members of his family. The workers were happy with him at Karshak Kshetra.

As a farmer, Parameshwar had his own principle to treat his workers as fellow farmers and the members of his family but not as laborers or coolies. He was completely in contrast with his relative, Suryasen in all respects.

Parameshwar never wished to be feudalistic. It was against his principle. He was a model farmer. He wished his only son Vishesh to be a model farmer only.

Vishesh was pursuing his studies at Shantiniketan. He was about to complete his studies at the nation-famous institute soon.

Parameshwar was waiting for his only son, Vishesh, to complete his studies.

While reading the newspaper one day, Parameshwar was lost in deep feelings:

"The farmer, the backbone of the nation, must be ideal and model and due respect should be given to him since he's the food-giver and life-giver... A farmer mustn't be like Suryasen."

"I must be different from the despotic Suryasen in all respects."

"My son, Vishesh mustn't be like Suryasen's son, Indrasen."

"The people at Indrapur were afraid of Suryasen and his son, Indrasen as they had dictatorial attitude. To go against them in any issue was very risky."

"All the virtuous people everywhere like me hate Suryasen and his son Indrasen."

"My son Vishesh stands before me soon as a good citizen... a good farmer."

"Time glides, holding responsibilities for life tides. Time in its incessant glide leads to the life tides, reflecting changes in it and its incessant flow. The institute sends out my son Vishesh, shaping him into a perfect man. I'm very confident of my son's success as a student."

* * *

2

All that took place in the farewell flashed upon the mental screen of Vishesh when he kept recalling the past especially the feelings of outgoing students including him. It flashed on the screen of his mental screen:

The students start assembling at the Tagore auditorium on the premises of the institute to bid farewell to us, the outgoing students, on the occasion that year.

There's a sweet mixture of tears and smiles reflecting in our faces. We, the out-going students, are mainly shedding tears for we're leaving our esteemed institute and beloved teachers.

We've a feeling that we're leaving the institute to have our own plans about our future career.

The dais is in glitters with the decking of flowers in rich beauty and bright lights in multi-variety lit on the dais.

The cushioned chairs are for the Chief Guest and the farewell programme organizers to address us, the students on the occasion.

It's very pleasant for us to glance at the decently decked dais.

The sweet music is going to delight us to our hearts' content.

The Chief Guest, famous in the region, is expected to grace the occasion within a few minutes and she comes onto the stage.

The President is in his seat when anchor has called him. The Chief Guest, Mrs. Vidya Jyothi, the Hon'ble Minister for Education in the Centre is in her seat. It is the time for the farewell function to start in a befitting manner while the staff and we are standing, clapping as a mark of respect for the guests.

Now the dais is ready for the function to start.

The Chief Guest and others on the stage are lighting the lamps as a mark of devotion to the Goddess of education.

We're standing as a mark of adoration for the goddess of education while they are lighting the lamps and respect for our nation while Vandemataram is going on in a graceful manner.

The principal is ready to present his report on the institute. We're giving a big round of applause whenever he tells the greatness of the institute:

I'm immensely happy that our institute has students with unsurpassed talents for their unrivalled success as per the principal's report.

The principal appreciates our, the outgoing students' extraordinary talents in our respective fields. I, in my heart, feel pride and honor for the astounding results with high marks in the university. I am successful in all kinds of activities.

The Government is pleased to upgrade it into Shanitinikethan deemed university from the ensuing academic year. I, on behalf of the institute, profusely thank the Government especially the hon'ble minister, graced the occasion amidst us for granting the status of deemed university to our nation-famous institute.

We go on claping in excess joy on the minister's according of the deemed university status to our nation-famous institute. Our happiness knows no bounds on hearing the happy news.

I'm happy that our institute has a special place for the achievements and accomplishments of its students in curricular and co-curricular activities: debates, group discussions, role-plays, elocution, essay writing and other competitions.

We, the students, have spirit and skills in extra-curricular activities too. He appreciates us for participating in many events of games and sports and winning prizes at all levels.

We, the students, clap for a long time in joy at getting awards and accolades.

The principal compliments us, the outgoing students especially me, Vishesh. My happiness knows no bounds.

All are proud of us, all the outgoing students with multisided interests and multifaceted talents.

I'm happy that we, the outgoing students, are going to settle in our respective careers and fields.

I feel sorry for leaving the institute. It bid farewell to me. It's inevitable on my part to leave the institute. They all accord a warm farewell to me with our classmates.

They're wishing us all the best...

I express thanks to the Principal in my heart.

The Principal concludes his report. We, the students, give a big round of applause for his detailed report.

It's very pleasant for all present to listen to the Institute's achievements.

The minister's happiness knows no bounds on listening to the report presented by the Principal.

All the members of the staff who taught us, the outgoing students that year and preceding years, are referring to our talents and merits with all smiles.

They're happy to come one after the other to express their profound feelings. They refer to the incidents in their service that year.

They say that it's very pleasant to teach the batch of ours, the outgoing students. I'm very happy indeed.

They feel pride and honour for teaching us.

We all feel very emotional when they recall their memorable association with us.

We all cheer by our clapping.

The Chief Guest comes to the podium on the dais amid the cheers and claps of us, the students. She starts to speak.

She's very happy to associate herself with the farewell function. She addresses the gathering with feelings of joy.

She wishes us all Good Morning... At the outset, she heartily congratulates us, the outgoing students, for the successful completion of our studies and achievements in the form of mementos.

She opines that every batch should leave an indelible impression on the institute like that of ours...We've played our role in that remarkably.

I as a student in the institute feel delighted to clap heartily.

She appreciates the outgoing students, especially me, Vishesh for my skills and talents, services, and successes. She wishes us all success.

Finally, she exhorts us, the outgoing students, to be models and ideal in life and distinguish the good from the bad only to learn the good and prove ourselves as good citizens.

I feel thrilled when she says that ours is the nation-famous institute and we are the students of principles to contribute to the progress and welfare of our nation that is facing unrest due to violence in the forms of rapes, killings, kidnaps, ragging and other types of violence.

Students today are the citizens of tomorrow. We as citizens feel that it is our prime responsibility to stop the incidents of violence and establish peace.

We should not shirk that responsibility since it's the utmost and a must.

We give a big round of applause on the part of all after listening to the address of the Chief Guest.

The Principal felicitates the minister in a befitting manner. Then she leaves, thanking all.

Now it's the time for us, the outgoing students, to speak about our life and associations with the respectable staff and other students.

First, I, Vishesh, come to the dais and speak, expressing my feelings on leaving my beloved institute, Shantiniketan and its members of the teaching faculty.

I'm Vishesh standing before you. I've a long association with the institute, Shantinikethan. I came here as a child to seek admission into the institute and studied from K.G to P.G. for eighteen long years. The thought of leaving this institute fills me with deep feelings. I'm very much indebted to the institute.

The teachers are living gods as they have seen and taught to shape us, all the outgoing students into successful students.

The communication skills I have today are because of you, my dear masters, Dr. Sandesh, Dr. Mahesh, Dr. Samatha, and Dr. Meena, seen in my presence. How can we forget you and your invaluable services?

The teachers at our graduate study excelled like PG faculty in teaching. They filled our minds with perfect knowledge in their subjects.

I can't forget the teachers at the plus-two level.

I remember my teachers at school. The knowledge given by them is still fresh in my mind. You've shaped me into a bright student.

I bow my head before you as a mark of respect to you, all my beloved teachers."

The students express their joys by clapping heartily.

Secondly, Khuresh comes forward to speak, expressing his heart-felt feelings,

He says," What we are today is surely because of our teachers and their devotion to our welfare. I can't control my emotions on the eve of this farewell function. We may forget anyone but not our parents and teachers. Parents give birth to us, and teachers give us education for our career in our lives."

The claps go on after Khuresh's speech

Divya says, coming to dais, "Dear Sirs and Madams, 'gururbrahma gurur vishnu, gurdevo maheshwarah'. I've studied here for eighteen years. Shantinikethan was our world...our bower to enjoy its beauty and gaiety. Today the thought of my leaving the institute fills my heart with tears. How can I forget my association with the esteemed institute and the revered faculty? It's impossible for us..."

All the students give a round of applause.

John says next, "Life is a ladder of two stages: the first stage of pursuing studies, aiming at the developing of one's career and the later stage of living a life in the career of one's choice. I bow my head to you, all my teachers as a mark of high reverence for you."

There is a big round of applause from us, all the students, after the emotional speeches of our faculty.

We, the students, feel very emotional about our intimate attachment with the teachers. We feel sorry for the feeling of leaving the sweet bower of their institute.

We humbly stand up and sing a chorus song, expressing our feelings. All clap in joy-grace accompanying our song-race:

Refrain: Shantinikethan, the bower for joy-showers

Our teachers offer all knowledge-treasures.

Stanza: It is only the address of love and affection

All teachers know teaching in full devotion.

Stanza: To celebrate all our merits is its tradition

All teachers teach in detail every portion.

Stanza: The institute of nation-famous recognition

A credit for the students sought admission.

Stanza: We come out of it singing in full jubilation

For name and fame in imparting education.

Refrain: Shantinikethan, the bower for joy-showers

Our teachers offer all knowledge-treasures.

The lyrical beauty of the chorus song with its musical quality is all for our gaiety in the farewell function.

There are cultural and other programmes. We dance and sing for our pleasure. We share the pleasure of the farewell programme.

The national anthem goes on and we all stand up to mark the conclusion of the farewell program that year.

We all have delicious and sumptuous meals with the items in variety while sweet music is going on. That's the function of ours, all the students and teachers participating in dining together.

We seek the blessings of our teachers, bowing our heads to them in humility.

It's memorable. We cherish the day in our lives.

The outgoing students come out of the institute, waving to our teachers humbly and filling our hearts with sweet memories in our minds until the last moment of our study at the institute, Shantinikethan. We leave the institute when our eyes well with tears to live our lives, adopting our careers as per our wishes.

....

While I was recalling all my experiences in the institute and pleasures in the farewell fuction. All the scenes of the function moved live in mind like a stream in its constant flow like that, the express train suddenly halted at a station.

The sun had already risen and was shooting its newborn rays directly to Vishesh.

Then Vishesh started to remember his parents: Parameshwar and Prasanna. His face started to glow in happiness.

Vishesh recalled his father, Parameshwar's aim to educate him at Shantiniketan throughout his journey. He thought of plans for his future to satisfy the wishes of his parents, especially his father.

His mind was busy planning his future in the advice of his father Parameshwar. He went on dreaming of bringing name and fame to his parents and to the nation by being a model in his profession.

Meanwhile, the station he was to get down arrived panting and puffing in the evening. He got off at the station.

His father Parameshwar came to receive Vishesh in a bullock cart. He was very happy to receive his son.

They got into the bullock cart. They went home, speaking to each other happily.

Vishesh enjoyed his journey to his native place in the bullock cart, watching the sights and enjoying the sounds of birds.

When the cart reached the house, Pasanna, his mother received her son Vishesh very happily. He was very happy in the company of his parents.

Vishesh felt comfortable at home by the time the sun was about to set on the western horizon amid the pretty barred clouds.

Meanwhile dinner was ready. Vishesh had dinner along with his parents and went to bed. He slept speaking to his parents in affectionate terms.

*****

3

Vishesh left the residential institute of his studies from KG to PG with deep feelings the day before yesterday. He travelled to his native place yesterday. He was at home. He was not able to forget his institute, Shantinikethan named after the Poet of the Universe, Rabindranath Tagore's Shantinikethan. It was a very famous institute for its aims and objectives in the welfare of its students. Even today all the people in the nation therefore prefer it to most other institutes for admitting their kids in the institute as it has created unrivalled history for several years.

In the early morning, Vishesh felt that he had lost the most precious thing, the paradise-bower of education. He felt that it was nothing, but Shantinikethan that filled his heart and mind with sweet memories and pleasurable experiences. It was the most valuable thing not stolen by thieves but recalled in times of leisure for unbound pleasure.

Vishesh fell in the mood of recollection about his institute, Shantinikethan for a while. He went back to the students in his recollection. He imagined how he joined it though he did not remember the stage of his kindergarten.

Parameshwar found the institute ideal and model for educating his son, Vishesh. He expected the institute to shape him into a perfect citizen. He wished his son to rise far above the average standards of the children in village and hamlet schools and so he sought admission in it for his son.

Shantinikethan was the place of education with knowledge and wisdom. It gave him all pleasure for it shaped him into a perfect citizen. He wished that he could be reborn to study there from K.G. to P.G. as he revered it as a model institute. It was the most reputed institute with the principal, and the teaching faculty and the academic atmosphere amid various trees.

Parameshwar, his father belonged to the clan of splendors. He was the owner of a hundred acres of land in the hamlet, Karshaka Kshethra. He had many workers to till his vast tracks of land. He built the hamlet with a hundred and odd houses for the shelter of his workers in the land. The people working on the land were uneducated and lived in the houses of the hamlet.

Parameshwar as an illustrious farmer supervised the works in the fields, living along with his workers in the hamlet, Karshaka Kshetra. He shared their joys and tears. He was their well-wisher to the core.

For the education of the children of workers, Parameshwar established a school at the hamlet. With the help of some scholars, he conducted an interview for the selection of worthy teachers for the school. The candidates who attended the interview were not up to the mark. He expected brilliant candidates to attend the interview, but it did not happen like that, as many were not ready to work as at a hamlet school. He was to select all the candidates who came for the interview to be teachers to work in the school.

Parameshshwar inspected the school and its standards from time to time. He was not fully happy with the instruction of the teachers. He advised them to improve the quality of their teaching. He expected the teachers to better the quality of their instruction and enrich the standards of the school in his supervision.

Apart from the school, Parameshwar thought of the health and well-being of his workers, "No body should suffer from any ailment." For that wish, he appointed doctors to work in the clinic that he built for the medical facilities of the people in the hamlet. The people were sure of the doctors to cure their common ailments.

Parameshwar provided all the facilities in the hamlet. The people felt it fully equipped and facilitated. It consequently became a model hamlet in the region in his care and concern, love, and supervision.

The Central Government heard about the village. Some officials visited the hamlet and surveyed its essential facilities. They were happy for their well-being. They awarded the best village award to reward the relentless efforts on the part of Parameshwar.

All in the hamlet, Karshaka Kshethra were happy with the facilities, but Parameshwar was not happy with the instruction in the school eighteen years ago, He, therefore admitted his son, Vishesh in the nation-famous institute, Shantinikethan, expecting it to shape him into a prefect citizen and ideal man. He wished him to have good instruction and achieve an illustrious career.

Like that, Vishesh fell in the mood of recollections until he heard a call.

Vishesh looked at his father Parameshwar when he called him for his response. He humbly responded to his father to be in the real world of the present from recalling the experiences in the institute and the life in the past. He was happy to speak to him. He expressed his feelings of joy on his arrival at his native place, Karshaka Kshetra,

"I'm here at your behest to speak to you...and listen to your words to the full."

"I congratulate you on the successful completion of your studies at Shantinikethan," said Parameshwar.

"Thank you very much, my dear father," said Vishesh with all smiles.

"I feel that we're two, you and me, to think of our workers' welfare. For me, the farmer, indirectly the worker, the cultivator of the land is the backbone of the nation. We're for the workers, my dear son," said Parameshwar.

"Thank you for shouldering the responsibility on my part. I'm proud of my father and his ideals in the welfare of the farmers," said Vishesh.

"We're for the workers. They're not just workers but farmers as per my principle. They're the partners in the industry of our agriculture," said Parameshwar.

As Parameshwar expected, Vishesh studied KG to PG and came out of the institute with unrivalled success. He was happy, as his son was a brilliant student of extraordinary talents. He understood his principles and ideals, aims and objectives as a farmer to be exemplary.

During his student days, Vishesh visited Karshak Kshetra very rarely. He stayed most of the time in the hostel and studied at Shantinikethan for eighteen long years. He did not know much about his father and his ideals when he stayed at the institute then.

After the completion of his studies, Vishesh started to live at the hamlet, Karshak Kshetra. He communicated with its people in a friendly manner, sharing their views. He found them very disciplined. They were very loyal and obedient to his father. They behaved and moved as per his father's instructions. They all respected him for his great qualities, as he was a farmer to all workers like a king to the people. He loved the workers as the people, the members of his family, for he was good at heart to treat them with love and affection. He was interested in their welfare and so he became a famous farmer in the nation. He was not just a farmer, but also a leader to lead them in their welfare. For him, the farmer is the leader to lead his workers well.

Vishesh was very happy in the hamlet for he found his father maintain cordial relations with the people to reflect human values. He expressed his joy over the hamlet's peace and harmony,

"Father, I'm happy for you love your people very much."

"Okay, my son, I expect you to have love for the people," said Parameshwar.

"I love them in the way you love them. They're my dear and near."

"I wish that you should be an ideal farmer to lead the people as an ideal leader for their welfare. At any rate, the people shouldn't be losers," said Parameshwar.

"I devote my full time in the welfare of our workers," said Vishesh

"That's my cherished dream. I've dreamed of my son having ideals like mine. That is why. I educated you at the model institute, Shantinikethan for eighteen years," said Parameshwar.

Behind Parameshwar's success, there was a woman, his wife, Prasanna.

Prasanna as wife of Parameshwar was ideal. She was cent percent ideal like him. She cooperated with him in all respects. She minded her household chores and familial duties. She advised him if at all she felt it necessary to advise him. There was no occasion for her to advise him, as he did not know the false ways. She was a perfect homemaker.

The irony was that Prasanna was the sister of a farmer-cum-leader, Suryasen of Indrapur. She did not like his dictatorial qualities.

Indrasen followed his father Suryasen in all respects. But Prasanna did not possess the qualities of her brother, Suryasen, nor did she have the qualities of her bother's son, Indrasen. She never liked them both in the family. She said sometimes,

"I haven't inherited the qualities of my brother Suryasen."

"I'm happy that I'm the wife of Parameshwar with all good qualities."

"I'm fortunate to have a son in the way his father feels and wishes."

Prasanna was not proud of her brother, Suryasen and his son Indrasen. She felt sorry to reveal their qualities to her husband, Prameshwar any time.

Prasanna was a different woman all together. She was an ardent devotee of Goddess Saraswati. She and her husband, Parameshwar had the darshan of Goddess Saraswati enshrined at the temple at Basara in Telangana. She prayed for the welfare of the people across the world,

"The people all over the world must be happy. The world should be free from violence. It should enjoy peace and harmony of the people. It shouldn't face any wars or battles on the borders or any untoward incident to disrupt its harmony."

"The people all over the world should enjoy wealth-like health. Health that is wealth is for strength. We've strength from food to be essentially available to us. For food, there must be food production. For food-production, there must be crop-production. For crops, there must be systematic farming for good yielding. For all that, there must be the farmer for well-farming and good food-production for the health and the wealth of the people across the world. The food for us to eat, clothes for us to wear and shelter for us to live are essential. We all should have all these. For all these, I need the blessings and benedictions of the goddess, Saraswati."

"The people must be educated as education is to lead us from darkness to light. People should have enlightenment by means of proper education. You, my Goddess, bestow on the people all these across the world."

"There shouldn't be killings, rapes, kidnaps, thefts, ragging, smuggling and the like of violence mustn't be in the world."

Prasanna was a woman who did not have parochial aims and selfish goals to live within the narrow walls of all barriers. She wished for the welfare of people across the world but not just hers or her family and relatives. Her love was for the race of man.

Who was worthier--Prasanna or Parameshwar? Neither was greater than the other. Neither was inferior to the other. They were complementary to each other. They were the ideal couple, and all felt them to be equal to be complimentary.

Vishesh was the only son of the ideal parents. The parents wished an ideal life for their son as man for the welfare of man's race.

Prameshwar and Prassanna lived happily with their only son, Vishesh. They liked their son very much. They wanted their son to treat their workers well in the way they treated their son.

4

It was the house of Suryasen, lavishly built for luxuries at Indrapur. He was different from her sister, Prasanna. He was a dictator, and she was an ideal woman, the wife of Parameshwar. She proved herself a model homemaker.

The irony was that Suryasen and Prasanna as brother and sister were not similar in nature. They had the goals in different ways.

Prasanna and Parameshwar proved to be models and ideal for they treated their workers as the members of their family. The family of Suryasen and Suprabha had a bad name for treating their workers as slaves. The two families of Parameshwar and Suryasen, diagonally opposed to each other in all respects. Suryasen's family, known for treating their workers as slaves, was living in neighboring villages.

Suryasen posed to be superior to others including his close relations, as he was a leader, elected by all false means. He did not care for his respectable brother-in-law, Parameshwar, and sister Prasanna.

In the house of Suryasen, there were many workers to do various works. Malli was the main worker. She cooked for the family. She looked after Indrasen, the only son in the family. She looked after the couple Suryasen and Suprabha too. She looked after their agriculture. She mainly looked after Suprabha.

Malli was very dutiful and so she worked very sincerely in the house. Whenever she had time, she supervised the work of the workers in the agricultural fields. To all intents and purposes, she was the manageress rather than a worker of the family. She was more like a member of that family.

Malli grew and lived as a member of Suryasen's family. She expressed her feelings for the family,

"I respect the family."

"As a child, I joined here to work as a servant-maid. The family nourished me from a poor family. I must wish the welfare of the family."

"They depend on me for all. They sometimes sought my advice. I feel honored for that."

"Once I depended on them for food and clothing."

"Now I'm the member in the family."

"I wish that all in the family should be well."

Those were the feelings of Malli for the members of the family. She joined the family as a child laborer. She grew to be very beautiful.

Malli looked different in the family, as she was a beautiful woman even in simple robes. When she grew to be in youthful exuberance, she stole the glances of all the people for her winsome features.

In the region, Malli surpassed the beauty of the wives of the zamindars, big landowners who were feudal in nature but never amiable and friendly as per human culture. They never appreciated the merits of others. They tried their best to excel her beauty but failed miserably. She wore simple saris but looked decent and dignified. In fact, the saris she wore were the old saris given to her by them. All grew to be very jealous of her unrivalled beauty.

Malli was in the height of five feet four inches. Her tallness was her special feature for her attraction. Her rivals with their jealous feelings posed to be tall like her. They tried their best to look tall by wearing high heel footwear. Sometimes, they fell off when they walked with high heel footwear and the people laughed them secretly.

The wife of Suryasen, Suprabha felt jealous of Malli's beauty. She sometimes was bound to express her feelings openly,

"God is very kind to you in creating you with special care."

"Why do you say so, my lord's wife, *Dorasani*," said Malli?

"See how I'm... See how you're," said the wife of Suryasen, Suprabha.

"How I'm, you can tell me?" said Malli.

"Don't you know about yourself?" said Suprabha looking at Malli from head to toe leisurely.

"I'm a servant-maid...That I know perfectly," said Malli.

"You're...You're...You're..."

"What...?"

"You've that I don't have," said Suprabha.

"What are those?" said Mall

"See I'm not slim like any cinema heroine. I'm hefty...I'm like an inflated balloon. I resemble a rice-filled jute bag. What should I tell you about myself?" said Suprabha looking keenly at Malli's charming features.

"You're the wife of my lord, Suryasen," said Malli.

"My hair is not quite long and jet-black for my hair-plait to swing in grace when I walk with my husband. It is very short like a broken tail," said Suprabha, looking at Malli's hair.

"I'm not tall to look graceful and beautiful...I'm short like the pumpkin," said Suprabha, raising her head to see Malli.

"You wanted to tell me about me...but you aren't telling about me, your servant-maid," said Malli to know about herself.

"I haven't yet completed telling you about myself," said Suprabha.

"Yes, please...You can tell...I have pleasure to listen to you attentively. Can I remain without listening to what my lord's wife attentively for long?" said Malli.

"If I tell you all about myself today... I don't have anything... at least something to tell you tomorrow, the day after and in the days to come... the months to come... the years to come... O! I want coffee," said Suprabha.

"Okay, I look forward to listening to you, my lord's wife in the days, the months and years to come... Right, I bring coffee to you soon," said Malli, rushing to the kitchen.

The significant quality of Malli was that she had qualities of head and heart. She had a keen perception and so she was very sharp in grasping all when she listened to them once. She heard the film songs on TV once or twice and sang in the same tune. She sang many songs at ease maintaining the correct tone and tenor.

Whenever there were no rains, Malli sang the devotional songs in chorus as her worship-offer to Varuna, the rain God for granting the shower of rains in the region and else where in the welfare of all farmers.

None was equal to Malli in her activeness. She actively participated in all the rituals of women especially in the celebration of Bathukamma, the most loved and revered festival in the Telangana region. On the eve of the festival, she dressed herself gracefully and walked with the cone shaped Batukamma of flowers in variety in her hands, all were looking at her in full fascination.

As a worker, she held the frame of Bathukamma in her hands as only servant-maids carried the flower frames to the place for playing it in devotion because the *dorasanis*, the wives of the feudals did not hold and carry them. They joined singing the rhythmical hymn in chorus and played together. Finally, they shared the sweets they brought with them after playing Bathukamma.

On the eve of Bathukamma, she was in her best sari. When all had impressions about her beauty, the women who grew jealous of her charms day-by-day felt more jealous than before.

The wife of Suryasen, Suprabha was very proud of having the beautiful worker, Malli in her house while others did not have a worker like her. Suprabha felt superior to others in that aspect. She posed for that, looking at her during the celebrations of Bathukamma.

Indrasen, the only son of Suryasen looked at Malli's unrivalled charms on the day of Bathukamma celebrations. He was infatuated with her beauty. For him, she looked more beautiful than before and so she drew his attention more and more all the time in her rich costumes on the day of Bathukamma celebrations. In fact, the costumes got more attraction when worn by her. He looked at her without turning his looks. She walked gracefully, excelling all others. She was the cynosure of all eyes on the occasion.

All the Bathukamma celebrations were going on in religious fervor in the place of their choice. Their playing of Bathukamma was coming to an end.

It was the time for all to leave the frames of Bathukamma in the flow of the village stream. Indrasen suddenly came to help Malli leave the

Bathukamma flower frame float in the stream. He took it and left the frame of Bathukamma flow in grace on the surface of flowing water.

Indrasen was in full appreciation of Malli's beauty and dreamed of enjoying it again and again thereafter.

The next morning Malli was in the kitchen, preparing meals. Indrasen entered the kitchen, as he was sure that his mother was sleeping in her bedroom. He entered the kitchen appreciating Malli's beauty,

"Malli, Day by day you are growing more beautiful...ravishingly more beautiful. I lavishly love you...My love for you knows it...I do not know how to express it. Yesterday you looked exceedingly beautiful in the Bathukamma celebrations. You were the centre of attraction since you looked more beautiful than all others in the celebrations."

"Young Lord, you looked at me, your servant-maid, with some affectionate concern and good looks as I work in your house sincerely," said Malli, bowing her head in humility.

"Not that... You really looked more beautiful than all others did though they were decked with scintillating ornaments and dressed in glittering saris. You were the cinosure of all eyes. All eyes fell on you...your beauty...I wish that you and your beauty must be mine...my wealth...my treasure," said Indrasen with all smiles.

"God had a lot of mercy on me to bless with the rare gift. In fact, God would have blessed the wives of the big farmers with my beautiful charms. By mistake, He blessed me with the charms," said Malli.

"Beauty is a boon rarely bestowed on women. You were God's best choice. He bestowed on you all beauty in plenty," said Indrasen, coming near her as if he had come to see the curry she was preparing.

"My beauty would have had rewards if I had been born rich," said Malli.

"I'm proud of you being in our house though not born rich," said Indrasen.

Indrasen came very close to Malli and kissed her, taking her into his arms for his deep hug with her. She had nothing to say except to experience all fears in her heart for the dictators, as she was after all a servant-maid working in the house for a long time. He kissed her again while holding her in his arms for a long time. He felt very much delighted for his union with her, but she was like a doll in his hands.

Malli became a beautiful doll in the hands of lusty Indrasen. She did not know what to do or where to go. She did not know how to react even in their union. She did not know that it was her yielding to Indrasen. If it was yielding, it might be due to her intimacy with him or had to be because of her innocence.

Indrasen was happy for his act and said while going out of the kitchen.

"I come... I come again...like the bee to hover around the flower and enjoy its beauty."

Indrasen met her in the kitchen everyday. One day his parents were away from the house as they went to watch a movie in the city. It was a movie recently released as the people talked about its merits to be a hit. When they left, the house was without his parents.

Malli's mind was filled with innumerable feelings. She did not know what to do.

Indrasen felt free and so rushed to Malli. He took her out of the kitchen, showering praises on her beauty. She was silent while he kissed her numberless times.

"I'm happy that heaven has descended on me into my house...into my arms...into my heart. God is great in creating you as beautiful as no other," said Indrasen.

Indrasen took her into his arms and had sex with her. He satiated his lust.

Malli deeply felt that she had lost some precious thing.

However, Malli was a plant in the garden of weeds. She was a beautiful flower to become a victim to heavy storm. She was a lovely youth within the reach of a lusty ruffian, Indrasen but she did not express her feeling to him, as he was her young lord of the house and the only son of a despotic leader, Suryasen. She did not speak any word against him for his act nor did she spell any word against any member of the family.

Malli proved that she was innocent, as she kept silent about the violent action. She did not know what to do. In Threta Yuga, Ravana abducted Sita for his evil goal. In Dwapara Yuga Dushyasana disrobed Droupadhi in the royal court in support of Suyodhana. The demons Ravana, Dushyasana and Suyodhan and the like reaped due consequences for their brutal actions. Indrasen raped innocent Malli like many others of barbaric nature in this Kali Yuga, the age full of evils and violence.

Indrasen frequented to meet Malli in the kitchen even when his parents were at home. The kitchen was only the meeting place for him with her.

In a happy mood, Indrasen sometimes felt like going out for films with Malli. The idea of going out with his servant-maid to a film became an obstacle for him. He repeatedly thought that she was his servant-maid, laborer, and worker. That was why he never wanted her to take her to a film or any function.

Malli thought that their union was because of their intimacy with the sense of care and concern for him but she never thought that her union was due to her being in youth.

Malli never had an idea of having any amorous feelings for Indrasen. She was busy minding her duties.

5

Parameshwar was ideal. He set a model to the people of Karshaka Kshetra as a farmer. He was for his workers. He loved his workers and wished their welfare forever. They also wished for his welfare. They were like members of a family. He treated his workers as their partners unlike many others who treated their workers as slaves because they were born low for low works like labor to work in the field or in the house.

Many of the landowners and leaders fell into false ways, as they were feudal with their whims and fancies to treat the people as untouchable and unrespectable. They robbed their own workers of their sweat. They exploited the innocent to meet their greed not to meet their just need.

The leaders aimed at success in elections by hook or crook. They supplied drinks to the voters to drink to the brink and offered notes for votes for victory in the elections. They enjoyed their powers as leaders. He never liked them for the clan prejudices or caste prejudices.

There was another quality that Parameshwar knew about many leaders, to treat their people as slaves. He even saw his own relatives treating the workers as slaves. They were Suryasen and his son Indrasen in the neighboring village, Indrapur. He was not happy with them. Suryasen was a dictator treating his people as slaves. His son, Indrasen was following his father's footsteps in treating the people as his slaves.

In such unwelcome situation in several places, Parameshwar and Prasanna prayed to God to bestow on their son, Vishesh all the good qualities of head and heart. He should have regard for his fellow beings and wish for their welfare.

The parents wished Vishesh to be free from all evil influence and the acts of violence to effect in tears. He should not have the influence of such leaders and fall into their false ways. He should be like the lotus leaf and not allow any drops of evils to touch it at any cost.

Parameshwar sometimes doubted that Vishesh, despite his having studied at the esteemed institute, Shantinikethan for eighteen years, might fall into the ways of such leaders. He wished to know whether he would fall into false ways or not. He sometimes recalled what many people expressed that the innocent fall into the false ways,

"Evil influences will definitely affect nature and the behavior of innocent people as they are easily tempted by evil forces at some weak moments. There may be some scope. We can't rule out that influence on the innocent."

"If a good boy lives with a bad fellow for some time, the good fellow will become a bad fellow when the good fellow is not careful."

"One's greatness lies in not being influenced by the bad and the vicious."

"You should remember what great people said, “Great people lay good roads to lead to great goals."

"One should be cautious about evil influences on the innocent."

"A foolish friend is not better than a wise enemy, but the former is friendly, and the latter is enemical. The former can influence others easily while the latter is silent. The former can teach all good people to learn his evils. They are likely to be temptated by the teachings, but they never think so. It appears to be fool-proof for them to be evil-free. It is a test on their part."

Keeping in view all these, Parameshwar still doubted whether there might be evil influences on Vishesh when there was a scope for him to fall in a bad company. He wanted to clarify his doubt and so he decided to ask his son to visit the house of his relatives, Suryasen and his son Indrasen in the neighboring village, Indrapur and stay long as per his wish.

When Parameshwar felt so, he said to his son,

"For a change, you can visit your uncle, Suryasen's house. There is his son, Indrasen in the house in the neighboring village, Indrapur."

"Okay, my dear father,"

Vishesh went to visit his relatives in the neighboring village. He got a simple reception at their house as they were in a different nature. He found a different atmosphere in the house to mark a difference from that in his house. He thought that he would not like them and so he would not stay in the house for a long time.

His uncle, Suryasen was a leader to represent their constituency earlier. He was more a dictator and feudal than a leader and representative for the people of his constituency. He was like a despotic king. He treated the people as slaves than as the people to enjoy their fundamental rights in the democratic domain.

His son, Indrasen, was more a prince and a feudal than the son of a public representative. He commanded his workers. It all appeared to

Vishesh that Indrasen was not fit to be a leader in the democratic set-up.

Vishesh found dictatorial attitude in the members of the house especially in his uncle Suryasen and his son Indrasen towards the people in the village. The people removed their footwear whenever they reached the gate of the house. They spoke to him with folded hands. Whenever his son, Indrasen called someone to attend his work immediately, he said in a commanding voice, "Who is there"? There rushed to him four to five workers to attend his works, saying,

"What do you want, my young lord? We are here."

"Where have you, all the fellows gone?" said Indrasen angrily.

"We're here only at your behest... at your disposal. Where can we go leaving you, my young lord?"

"Go and bring the newspapers from the drawing room."

Three servants ran one after the other to bring the newspapers. Meanwhile he felt like having coffee. He said loudly, "Where is coffee?"

Vishesh, who was with him, observing his commands and demands, felt astonished at his dealing with the servants and others in the family.

Three people who had rushed to the drawing room brought the newspapers and put them in Indrasen's hands. He started to go through the papers anxiously to study the pulses of the people in voting.

"Vishesh, I'll contest the election this time also. To file nomination, there is forty days time. Elections are near not after long time. They're near to think of them seriously. I want to contest the election. Last time I got thumping majority. This time, I'm sure of getting more votes if not the same majority the last year," said Indrasen to Vishesh, posing.

"I appreciate you for winning the hearts of the people to get the land-side victory in the election," said Vishesh.

"That's Indrasen... Indrasen's capacity...Indrasen's popularity, Indrasen's powers, Indrasen's strength," said Indrasen.

Indrasen did not tell Vishesh how he had won the election with a thumping majority. He did not refer to the false ways of spending money and supplying drinks to the voters to get votes.

Meanwhile, two cups of coffee came. They had coffee while Indrasen narrated his family's long story of elections to Vishesh that his father, Suryasen was the representative of the constituency for two decades earlier. The people knew him as a leader of success.

Indrasen called a worker to find out about the completion of a work he assigned to him two days ago. Then three workers came one after the other to him,

"What happened to the work I assigned two days ago?"

"The work is going on, my young lord," said the worker and other workers nodded their heads with folded hands.

"Are you playing there without working? It's the work of two days."

"But..."

"No but...Say 'yes' to the completion of the work as per the time given to you for its completion. Take a day to complete it. If it isn't completed in a day, I will beat you like anything...I break your bones..."

"Young Lord, we complete it in a day."

"Workers, get lost. Show your face when you complete the work in a day."

All the three workers disappeared one after the other with folded hands. They shut their mouths. They were not able to express the reason for not completion of the work. In fact, there was disturbance in work due to inclement weather.

Vishesh felt stunned at Indrasen's attitude towards the workers in one way. In the other way, he thought that the workers might be lazy. They might have neglected the work. Deliberate negligence might be there on the part of workers. The warning might be necessary in that case. He was not able to make out the real reason, as he had not encountered such incidents, faced by him or by his father in his village.

Vishesh was in the house of Indrasen for a fortnight, observing them and their behavior with their people. All influenced him slowly against his aims. His attitude towards people started to undergo gradual transformation against the principles he got at Shanti-nikethan institute.

During their conversation, Indrasen presented a clear picture of their clan's high dignity and vanity, spendors and powers. When Indrasen told all Vishesh learnt all those secrets of their clan.

"Vishesh, you and I belong to the clan of high splendors. We're like princes and our parents are like kings. Our parents and ancestors enjoyed high regard and rare honor of the people," said Indrasen.

"I see. I haven't known all these so far," said Vishesh.

"We should treat people as our slaves. It's our special privilege."

"I see."

"We're far superior to the people in all respects as we were born of the parents with highs and flairs."

Vishesh learnt from Indrasen about their clan and caste with special powers and rare splendors. He was to express his view to Indrasen as per the principle, "The people are born equal," but he did not reply so. He wanted to know Indrasen's responxe.

"I see...," Vishesh said, welcoming Indrasen's dictatorial feelings.

"When we treat them equally, how can we be in the dominant and commanding position and enjoy the superior status? To enjoy superiority and supremacy, we must treat them as our slaves. We can enjoy their services as slaves."

"O, we can treat the people as slaves for our comforts and luxuries," said Vishesh without hesitation.

Vishesh learnt from Indrasen about his clan and caste with special powers and rare splendors. During his stay for a fortnight with Indrasen, Vishesh's attitude towards the people started to undergo transformation in Indrasen's company.

Vishesh went back to his father's place, Karshak Kshetra with transformation in his attitude towards people. Primarily, he was an idealist as per the principles as he learnt allgood things at Shantinikethan as astudent. As a pass-out of Shantinikethan also, he did not know the way where a man was to discriminate his fellow being from him. Under the influence of Indrasen, there was a clear transformation in Vishesh's attitude towards people.

When Vishesh came back, his father, Parameshwar did not ask him to know what happened to him in the neighboring village, Indrapur. He found him thinking of something serious. He tried to know about the things that he did not like in the house of his relatives.

One day, Vishesh ordered the workers of Karshak Kshetra to do some small works like Indrasen. Parameshwar did not like the way he ordered the workers. He warned him that he should not order the workers like that. We should ask them to do the things with all love and affection.

6

One day, Vishesh declared his readiness to contest the ensuing election to his father. His father Parameshwar felt stunned for the sudden decision of his son to be in the fray like Indrasen.

"I hope Indrasen has already submitted his nomination papers. I also submit my nomination papers," said Vishesh.

Parameshwar understood his son's firm decision to file nomination papers and so he tried his best to convince him,

"Dear son, don't join politics and spoil your career. It's not your field. You've false temptations in it against your wish. I've many expectations about you. You're to fulfill them without fail."

"I enter the arena of politics,"

"You're for our workers after me. Their welfare is main for me and you," said Parameshwar to convince his son.

"I contest the election like Indrasen," said Vishesh.

"The political field isn't fair especially in modern times," said Parameswar.

"My conscience prompts me to be in the fray," said Vishesh.

"When I wish that you should be a farmer, a model farmer like me. You and me...we're for the farmers...for the welfare of all farmers," said Parameshwar.

"I enjoy powers more as a political leader than I do as a farmer," said Vishesh.

"You needn't be a leader. You can and should be a farmer to enjoy your life. The farmer is the real leader," said Parameshwar.

"I feel that the leader in the political field has more powers. He is in the commanding position," said Vishesh.

"A farmer is the bread-giver. He is the life-giver. He can serve the people by giving them food as the backbone of the nation. That is the real service," said Parameshwar.

"We can serve the people more as a political leader than as a farmer," said Vishesh confidently.

Parameshwar understood his son Vishesh's attitude towards the farmer to do agriculture. He felt sorry for his son who underrated the farmer. He never expected that Vishesh would join politics against his will and wish. He tried his best to divert his fascination for the spell of politics.

Despite his father's strong wish, Vishesh determined to join politics. The news spread everywhere and reached the ears of Indrasen as a surprise.

Vishesh wanted to start an office in the city nearby as he thought it would be very convenient for him to have rapport with many people. He took some houses for his office. For that, he shifted to the city to chalk out all plans for winning the election.

Indrasen filed nomination papers well in advance though there was still time for filing nomination papers. Vishesh was not in a hurry to file his nomination papers. Before the last date, he was determined to file his nomination papers.

When Indrasen heard the news that Vishesh would contest the forthcoming elections from the same constituency as his rival, he felt that Vishesh had superior qualities in higher education, enormous knowledge, handsome personality, pleasing manners and so on to impress the people. All his merits proved him worthy to be a public representative.

Indrasen was sure of Vishesh's winning the election. He seriously thought of how to keep Vishesh away from the fray.

When Indrasen did not want Vishesh to contest the election, he secretly plotted to keep him away from the fray as a rival to him. He thought of ways to see that he would not contest the election.

When the people heard about Vishesh's decision to contest the election, the news in newspapers and messages through WhatsApp and other social media went viral in Vishesh's favor, "Indrasen will taste the debacle in the forthcoming elections for his faults. Although he supplies notes to every man and woman and drinks for them to drink to the brink for votes, he would lose the election. Vishesh will win. It's sure and certain."

Indrasen seriously thought of the news and messages in the social media and evolved plans to keep Vishesh's name away from the fray. He decided to distract Vishesh's attention from the dream of becoming a public representative and contesting the ensuing election.

Through newspapers, Indrasen knew that Vishesh opened an office for elections campaign in the city nearby. He studied all Vishesh's efforts in becoming close to the people, the voters in the constituency.

Vishesh thought that he would become very close to the people. His habit of taking a walk would indirectly help him influence the walkers in one way. He took a walk in the morning in the Brindavan gardens of the city. He found many men and more women coming for a walk.

Of all the walkers, Vishesh was the most impressive person for his handsome features and charming manners. He was in the attraction of beautiful women. Indrasen, his opponent thought of him in many ways,

"Vishesh is in robust youth. He'll be under the attraction of a beautiful woman I send to him. She can easily tempt him with her unrivalled charms. It's certain. As a youth, I was also in the temptation of a beautiful woman. She yielded to my lust. She also easily tempted me. On those lines, the beautiful woman will surely tempt him by her lovely features. My plan will definitely work out on Vishesh."

To make his plot successful, Indrasen sent the beautiful woman, Laila to Vishesh, telling her where she was to meet him, how she was to tempt him by her charms, how she was to love him to turn him a slave to her love. He encouraged her to use all the ways essential to infatuate him without fail. He paid her a fabulous amount far above her expectations, promising to pay her twice as much on the successful execution of the plan.

Vishesh was in the Brindavan garden, taking a stroll as usual. Laila, sent by Indrasen, entered the garden for her stroll. She was taking rounds while humming a famous love song in a melodious tone. Her song drew his attention to it. She moreover walked lovely in the swan-grace like an angel par excellence. Her charms were unrivalled to infatuate him. She shone like lightning amidst dark clouds.

Laila created a magic of beauty to mesmerize Vishesh, keeping him spellbound. He absorbed and watched her without turning his glances.

Laila looked at beautiful flowers and appeared to speak to them with all love and affection for their floral grace. He forgot to file the nomination paper and canvass for his election. He was under the infatuation of her physical charms. Ultimately, he was in the hooks of her charming looks.

Vishesh was very handsome. He was in robust youth. He had a passion for making friendship with women of winsome features. Laila was beautiful enough to tempt him. He saw her and fell in love with her at first sight. One day, they met, and they met everyday in the garden for lovemaking.

In the infatuation of Laila, Vishesh not only forgot to file his nomination papers for the election but also his father who was the whole world for him. He dreamed of Laila every moment as he was in the shackles of her infatuation. He was not able to come out of it nor did he try to do so. They were very eager to see each other,

"I love you, my dear," said Vishesh with all smiles.

"I too love you, my dear," said Laila to Vishesh.

They met every day and expressed their love for each other. They spent all their time in lovemaking. They did not bear without seeing each other.

Love became their world. They were in the nest of love with them to enjoy it forever. They were the lovebirds in the air of freedom.

During their parleys, Laila expressed her special views, surprising Vishesh,

"I will stay with you for some time as long as I want to stay with you for my love to you. Later I go back to my region where I enjoy full freedom."

Vishesh was surprised at Laila's views as he expected her to stay with him life-long. He thought that she was in love with him, and he was in love with her. Love dwelled in their hearts like honey at the hearts of flowers.

As a true lover, Vishesh was sure that his love with Laila would not know any break. He thought that the necessity to feel heart-broken would not arise in his case, as theirs was not a temporary love-relation and affection-concern in their union.

Vishesh was sure of his success in love with her. He was very optimistic about his love for her. He expressed all positive feelings:

"I've never noticed any ill-feelings in Laila's face so far.'

"Our love is the garden to bloom flowers to spread its perfume of affection forever...Our love is an orchard to have fruits ripe for tactile bliss forever."

"Our love is a cloud for the showers of pleasures forever."

"Love knows no deception...no exploitation...no selfish goals."

"Love is flawless. It can never be lawless for it knows only purity."

"Love knows no break. It lives longer in the hearts of the lovers."

"Love is not selfish and one-sided. It's unselfish to be mutual and reciprocal...There's always reciprocity. It's love's life…life's love."

"I'm sincere in my love. The journey of my love is for love alone."

"All is well that ends well."

"My love started at first sight to live in our hearts safe forevr."

"My heart is pure. The love in my heart is like the lotus to let no drop of poisonous evil touch its heart... No evil will hurt its heart."

"I'm successful in my love like a brave soldier to win the battle."

"I've victory in my love like a hardworking student to come out with flying colours."

"Whoever is devoted to love will reap the fruits of love...the success of love... the rewards of love... It's the truth to reflect the beauty of love for the gaiety of lovers...That was my firm conviction."

Time flew witnessing happiness in the glows of their faces. Their days and nights went on in the might of delight. The birds on the branches of trees knew that they were true lovers. Sometimes the birds were jealous of them as lovers.

The people who were taking rounds in Brindavan Garden looked at Vishesh and Laila with rapt attention as they were in good impressions. All thought of them,

"We've seen many pairs of lovers, but this pair is with very winsome features."

"In case of the pair, it is very difficult to say that in charms the lover excels the beloved or the beloved excels the lover."

"As a couple, the lover and beloved are complementary to each other for the compliments of all who watch them from close quarters, as one is not happy without the other."

"Vishesh and Laila are the model lovers like Laila and Majnu."

"The lovers in separation and frustration should learn from Vishesh and Laila as in the case of the model lovers, Laila and Majnu."

"Selfish lovers get separated in haste, turning their love waste,"

"They're worthy to be the couple to lead their happy married life lifelong."

The walkers of Brindavan garden considered the lovers true lovers. They thought that they would not face any hurdles in their love.

....

It was a great pleasure for Dakshinya Deekshith to take the abandoned babies to the orphanage he built at the outskirts of the forest.

He made up his mind to adopt them on auspicious occasions. He found more female babies than male babies. He got male babies less as only most of the mothers were reluctant to abandon them. That is why he got a few male babies and more female babies.

His wife Annapurna cooperated with Dakshinya Deekshith in making his goal of adopting the abandoned babies a success.

Dakshinya Deeksith did not like mothers to abandon their babies irrespective of the reasons they furnished. He read in the newspapers that some mothers abandoned their babies in drains, bushes, wells and so on. The mothers were ready to kill the female babies by abandoning them. He felt that it was the worst violence and the most sinful act on their part. He felt pity and thought of a solution.

One day Dakshinya Deeksith seriously thought of the inhuman act of abandoning babies by mothers. He found a way out as his mission to give life to the abandoned babies. He thought of starting an orphanage on the outskirts of the forest nearby. He wanted to build it in the lap of nature. The land for the orphanage was near the forest with beautiful waterfalls. He adopted the abandoned babies and brought up them as his children. They grew up nearer to nature. He felt that it was the most congenial place for the orphanage.

To fulfill his mission, he made up his mind to keep all the abandoned babies in the orphanage although Dakshinya Deekshit had two daughters of his own.

7

In the days to come, Malli learnt that she had become pregnant. Her poor parents knew the reason for her pregnancy. They had much fear for she became pregnant without marriage on her part.

Malli's parents did not know what to do. They somehow wanted a solution to their daughter. They had innumerable feelings:

"Suryasen mayn't listen to our requests to do justice."

"Suryasen has no soft corner for the workers for he treats them as slaves."

"When we know his despotic nature, how can we approach him to have mercy on us, the worker's parents?"

"Let's go to his wife, Suprabha. She mayn't consider our requests. Suryasen might be angry with us for our approaching his wife."

"Let's go to him with folded hands for the justice done to my daughter Malli who had worked long in his house with commitment."

The parents of Malli approached Suryasen as per their decision and started to go to him one day. While they were coming near his house, they removed the footwear and dared to see him in his palace. They told him what had happened to their daughter, Malli with folded hands.

That was the time of the elections. Canvassing for the elections was going on. Indrasen had already filed his nomination papers and was in the fray. He spent a lot of money and supplied drinks to all the people in the constituency to be victorious in the election. He was sure of his victory with a thumping majority. He kept studying his son's winning chances on the elections,

"If the issue of Malli is raised, it'll definitely reflect adversely on the chances of Indrasen's winning the election."

Suryasen did not express any feelings to the parents regarding Malli's pregnancy. He appeared to listen to it for due attention to it.

"Let's wait for some time to discuss this issue," said Suryasen.

"Don't spoil the life of my daughter," said Malli's parents with folded hands.

"Haven't you listened to my words...? Are you deaf? I've already told you to have patience for some time... Leave the place immediately, shutting your mouth," said Suryasen angrily.

Suryasen was such a dictator that he did not care for anyone with due respect or concern. He was lusty. He had five concubines in the region. He secretly maintained relations with them. His wife, Suprabha did not know all that.

As it was the time for canvassing for the elections, Suryasen was busy. His supporters were also busy canvassing in favor of his son's candidature for his sure election. They were extremely happy for his son being in the fray. They welcomed his son's success.

Malli's parents wanted to inform Suryasen's supporters. They did not care for them when they approached them. They also said,

"Wait for some time. We're busy now."

Suryasen's supporters were not ready to listen to Malli's parents as they were busy, canvassing for the victory of Indrasen.

From inside the house, Suprabha, Suryasen's wife listened to what Malli's parents had told her husband. The family felt it would be a blow to the family at the time of elections.

Malli's parents wanted to ask their son, Indrasen but they did not see him. All of Suryasen's family kept quiet. They kept the issue of Malli as a top secret for it would spoil Indrasen's chances of winning the elections.

Suryasen did not call the parents of Malli to him to settle the issue later. He deliberately kept the issue aside. He indirectly wanted to close the issue once and for all.

To be away from the situation and make Malli's parents keep quiet, Suryasen thought of settling a match for his son, Indresen. He selected a match from the city nearby. She was the only daughter of a business magnate.

The city girl's parents had a good impression of Indrasen as he was a leader from political background, and he was in the fray that year. They felt its prestige.

It was the right time for Suryasen to perform his son, Indrasen's marriage in a function hall in the city.

The day came for Suryasen to perform his son, Indrasen's marriage in the city function hall. Many guests attended the marriage. After the marriage, he brought the newly married couple, Sukanya and Indrasen. He thought that the news of his son's marriage would silence the parents of Malli and their supporters.

Malli's parents spoke to their daughter in an amiable and amicable way,

"Dear Malli, the people are not sincere. They're ready to deceive others without any hesitation. In such circumstances, there's only way."

"What's that way?" said Malli.

"I advise you to go for abortion...That's the only alternative," said the parents.

"I don't go for abortion,"

"You can marry a boy of your choice...," said the parents.

"I'm not ready to go for abortion and kill my child nor am I ready to marry a boy to spoil his life."

"Didn't Indrasen spoil your life?" said her parents.

"When Indrasen spoilt my life, I can't spoil the boy's life. It's not good on my part," said Malli, shedding tears.

"Listen to us, Malli," said the parents.

"Never...," said Malli.

"Many innocent women like Tess as described in Thomas Hardy's novel, *Tess of the Dubervíles* became victims to lusty people," said the parents.

"I became a victim to lusty Indrasen in innocence. Now my war is against lusty people. I'm against dust-like lust and dirt-like flirtation. Hereafter the innocent women should not become victims to such lusty fellows," said Malli, crying in hoarse.

"Your experience must be useful to the people in innocence," said the parents.

"I don't want to go for abortion and let my baby, fetis in the womb, die due to my inhuman act at any cost," said Malli.

"We too know that it's an inhuman act."

"I start a movement in the welfare of women. I enlighten them not to fall victims like me to lusty people. There're many incidents of victimization on the part of females of any age. The people witness and forget the incidents so easily. As a victim, I can never forget," said Malli.

Her parents understood the firm decision of Malli and referred the issue to their supporters and near and dear.

"Proceed as Malli wants you," said their supporters to Malli's parents.

Then Malli's parents with some people went to Suryasen and raised the Malli's issue. He immediately called the police to foil them.

Suryasen called Malli's parents inside his house and expressed his anger to them,

"I've already told you to wait for some time."

"I heard that you had performed your son, Indrasen's marriage. Is it good on your part?" said Malli's parents.

"I shoot you if you talk of the issue of Malli now. When you're shot dead, you can't speak everything as you have done now," said Suryasen very angrily.

"You can shoot me before you shoot my parents," said Malli in anger and fear.

Malli's parents prevented Malli, telling her, "We've our ways to deal with them to find justice done to you."

The news about Malli's pregnacy went viral. Suryasen wanted to keep silent in the matter. It became the talk of the day in every circle and in every house though it was a top secret, open for all.

All the people learnt about Indrasen's deceiving the innocent Malli. As a result, she became pregnant. They took it as a serious issue.

All the people in support of Malli made up their minds to go to any extent to find a solution to the issue of Malli.

Malli gathered women in the village to tell them what happened to her,

"My innocence turned me a victim to lusty Indrasen. Nobody should become a victim like me. You must be very careful with lusty men."

"There're females of any age are becoming victims to rapes...gang rapes. When the rapists find the rape of their own people, they come to know how rapes will result."

"There're dowry deaths. We should teach a bitter lesson to dowry mongers."

"The servant-maids are meant for work and the due salary for their work. The dirty fellows make use of them to satiate their lust. When it comes to marrying the victims, the caste matters much, and the status comes in the way. To satiate their lust, their sociological status does not get in the way."

"The women are abandoning their babies in dustbins, drains, bushes and so on. How the babies suffer and die, mothers do not bother, as they want to be free from humiliation. Though they bother, it is of no use. Their bother goes to waste. It's murdering the babies by mothers."

"Women in case of pregnancy go for abortion. It's killing the fetus, the baby in the womb. It's infanticide... a kind of murder."

"This enlightenment is necessary on the part of women."

"The females are becoming victims in many ways. We should find a way to put a full-stop to such incidents."

"Hereafter, we should have unity for the cause of women."

All women started to sloganeer to vent their anger. They were ready to extend their support to Malli.

They sloganeered against the deception of Indrasen and atrocities of Suryasen. They decided to teach them a suitable lesson.

All the people in the village, Indrapur supported Malli for the deception done to her. They were not courageous enough to go against Suryasen, as he was a strong leader and dictator. The incident of Indrasen's deceiving Malli made them united as they had sympathy for her.

All united to see that justice was done to Malli. They went on strike before Suryasen's house, demanding justice for the innocent, Malli. Suryasen informed the police to come to the village. The police started in the van with guns in their hands. The people were shouting with their slogans:

"We want justice done to Malli."

"The criminal Indrasen must be under arrest."

"Atrocities of leaders and feudal farmers must come to end."

"Save villages from dictatorial leaders."

"The deceiver must be hanged in the presence of the public."

"We want justice done to Malli."

"What would you do if your daughter were in Malli's plight?"

"We're also human beings like you."

"To deceive Malli is inhuman."

"We want justice done to Malli."

The police came from the city and foiled the striking mob, gathered to demand justice done in Malli's case. Some ran panic-stricken at the sight of the police. Others stood there sloganeering in the ways possible.

The police foiled the striking mob. They threatened the people by saying,

"We're ready to shoot you if you resort to sloganeering. Your slogans won't serve the purpose. Go home to save your lives or you are shot dead like birds."

"We want justice done to the innocent woman, Malli."

Some how the police managed to disturb the mob as the people panicked for the threat of shooting, given by the police.

Suryasen and Indrasen were inside the house. Suryasen thought that the people would beat Indrasen who was responsible for the deception done to Malli. He was in a room with the doors locked from inside. They took instantaneous measures for their protection.

Within twenty-five days, the polling was due. The results were due for the announcement of results within two days after the elections. The Election officer was to declare the results of elections after twenty-seven days.

The news about Indrasen's deception for Malli went viral. There were posters seen everywhere stating Indrasen's deceiving Malli. All thought that Indrasen would surely face the defeat. They were sure of his debacle.

Suryasen and Indrasen did not stop their canvassing to win the election. They supplied notes and drinks to capture votes.

All the people in the region including the family of Parameshwar, Prasanna, and their son, Vishesh heard the news about Indrasen and

Malli. By that time, all were aware of Indrasen's deceiving Malli. In his deception, Malli became pregnant. For that, they felt sorry for Milli's pathetic plight.

Currency notes at the rate of five thousand rupees per voter and five bottles of various types of drinks reached every house. Their supporters did door-to-door campaign without stopping.

Suryasen was sure of his son, Indrasen to win the election. He made up his mind to teach a bitter lesson to the people who supported Malli's cause.

Although Suryasen and Indrasen did not come out of the house for fear of the people who supported Malli, their supporters made all possible efforts for Indrasen's success in the election. They begged the people to vote for him. The people on the other side wanted justice done to Malli soon.

In such circumstances, their supporters tried to keep the people in the dark in respect of Indrasen's deceiving Malli, saying in every house,

"Some evil forces and enemy elements are trying to defame Indrasen by telling all untrue and false for defeating him,"

There was a new candidate in the fray. He looked at the Samaritan to the core. He had unique approach in appealing to the people for a chance accorded to him for serving them with devotion to prove his sincerity.

The new candidate was Vijay with a high education and good culture. He had cordial relations and keen concerns for the people and their welfare. He was sincere in his approach and made his manifesto clear to the people.

"I'm for quality education for all, the development of the key industry of agriculture, good medical facilities in all the places. For that, I've entered the arena of politics for right governance. I'm for the welfare of the people. I give special attention to the needy and poor. You know what the previous leaders had done for you. I'm unlike them. Please give me a chance to serve you, my dear people," said Vijay with folded hands.

There were processions in every place in support of the candidate, Vijay's election, sloganeering,

"Democracy is still alive. If democracy is sick and dead, give life to it by your valuable votes. "

"Votes aren't just ballot papers but they are the heartthrobs of democracy."

"Votes are the wealth of the people, duly cast by the people and the treasure for the people."

"Votes are the lives of the people as they play a vital role in the lives of the people like you."

"You should be wise in selection of your leader."

"Never yield to money and drinks"

"The people are wise when the leader is wise."

"Dictators dead, democrats reborn"

"Good days have come for the people."

"I protest against Indrasen to have deceived Malli."

"I'm at your service in the welfare of the people."

"Think twice before you vote once."

"People shouldn't become slaves as they are only king makers and leader-makers in their welfare."

"Cast your vote for your Vijay."

Vijay approached every voter personally and told him or her about his commitment as a leader.

The canvassing for the elections came to an end. They knew that money went every door and drinks flowed for them to drink to the brink. They were in the mood of intoxication.

The people used their brains. They thought of what Vijay promised them and his principles before they enfranchised their vote right on the day of poll. Polling went on in a peaceful manner without many untoward disturbances.

The date for the poll came and the people enfranchised their vote-right to the maximum. The poll commenced and ended without any untoward incidents. As per the date for the declaration of results, the Election officer announced results in a peaceful atmosphere.

After the counting of votes on the counting day, the election officer declared the new candidate in the fray. Mr. Vijay won the election with a thumping majority. Indrasen lost his deposit as shocking news for him and his supporters.

Indrasen tasted the debacle even after spending crores and crores and supplying drinks and drinks to the voters to drink to the brink. The new candidate Vijay got a landslide victory in the elections.

Indrasen did not want Vishesh to be in the fray as he thought that he was a worthy candidate with winning chances. He deceived Vishesh by means of his plot but fell into the ditch he had dug for himself.

* * *

8

Vishesh fell in love with Laila. She wanted a conditional marriage, but he thought that she was deeply in love with him. He had full faith in her that she would live with him as his beloved and his life partner in life. He believed that the marriage was life long as they were in deep love.

Vishessh and Laila as lovers were happy, saying to each other that they were born for each other and made for each other. The happiness born in their love knew no bounds. They were on top of the world.

Vishesh never guessed his love and his marriage with Laila would be conditional. He believed that love is eternal but not conditional.

Time rolled in its course to show varied ways to Vishesh and Laila to lead their lives in different ways. To Vishesh's surprise, Laila was on her way to the region of freedom, leaving him for his life.

Vishesh woke up daily and looked at her in bed at daybreak. To his shock, one day, he did not find her by his side at daybreak. She left a note before departing to her region unknown and untraced.

Tears welled in Vishesh's eyes. He noticed the letter left by her. He found the same letter on his mobile. He was not able to read the note as his heart was throbbing. The letters in the note were glaring as he had layers of tears in his eyes. He was not able to the letter at a time due to the layers of tears. There were breaks in reading the letter. It took a lot of time for him to read the letter like that. She wrote, sketching her feelings,

"I was in love ... with you for some time... as per my promise to Indrasen. He mainly... wanted me to distract... your attention... in filing your nomination papers... for the ensuing elections. That was my promise given to him... on his paying of the fabulous amount to me... He further promised to pay twice... the amount he paid first for the plot. Very soon, I'll get... the amount. I'm going... to my place with my advice... to you not to search for me... I'm going to... my own region... to live in freedom."

Vishesh could not believe all those happenings to be against his wish. He read it once again while tears were still rolling down his cheeks. He was able to read its contents against his heart's content, but the words were still glaring when he read again,

"I was in love with you for some time... as per my promise to Indrasen. He mainly wanted me to distract... your attention in filing your nomination papers... for the ensuing elections. That was my promise given to him on his paying of the fabulous amount to me... He promised to pay twice the amount he paid first for the plot... Very soon, I'll get it. I'm going to my place with my advice to you not to search for me. I'm going to the region to live in freedom."

The tears continued to flow down his cheeks constantly. The cheeks that had glowed with the dimples were experiencing the melancholic flow of tears. He fell into the ocean of woes and throes, as his love with her broke in that manner. He was not able to forget their love for their happy life. She thought that her leaving him marked the end of their love, voicing a sad note.

:

Laila did not mind leaving Vishesh even after having lived with him and become pregnant. She did not continue to live with him further. He as a true lover thought that he had a lot of trust and that she would not leave him at any cost.

Vishesh felt separated from his beloved like the dove separated from its mate in a storm. He believed that lovers never know storms or cyclones to punish them by separating one from the other.

After the lapse of time, Laila delivered a pretty baby. She did not keep the baby with her, nor did she keep the baby with Vishesh. She decided to abandon the baby near an orphanage in the lap of nature.

As per her decision, Laila was on her way to abandon her baby near the orphanage and go to her region of freedom. She wanted to leave it in the lap of nature. Before abandoning her lovely baby, she expressed her deep feelings:

"O My dear... my life...the life of my life. I can't abandon you here but I'm abandoning you."

"My child, excuse me for abandoning you."

"I've no right to abandon you since you're my daughter...lovely daughter. You've every right to live with me, your mother."

"I love you in my heart...You live in my heart."

"I kiss you. I cherish the memory of your feather-soft tender kiss in my life. I live in your sweet memories."

"Mother is a mother... daughter is a daughter... Ours is the mother-daughter relationship but...I'm sorry to abandon you, my dear baby."

"I'm not an ordinary woman. I'm an extraordinary woman with super-human qualities. I live in my region in full freedom with my own people,"

"You're so smart that I can't compete with you in any way, my dear. You're beyond comparison."

"I'm unfortunate as a mother. I'm not fated to fondle you. I've no opportunity to see you toddling and lisping in a pretty way. I miss to glimpse at your tender smile. I miss to fondle you. I don't have that fortune. "

"In spite my abundant love for you, I'm abandoning you."

"Pardon me to abandon you. I wish I should revisit you again."

Laila looked in all directions. She noticed an elderly person with his two disciples taking rounds near the orphanage. After kissing the baby once again, she abandoned it in verdant nature beneath a bush near the orphanage. She felt that it was insulting for her to abandon her in the lap of nature.

Laila expected that the elderly man with his disciples would surely pick up the baby and groom it fondly.

The person was none other than Dakshinya Deekshith. He had come along with his two disciples to take rounds amid the flora and fauna in verdant nature near his orphanage. In his last round, he suddenly heard the cry of a baby beneath a bush.

Dakshinya Deekshith asked his disciples to go in different directions to search for the crying baby. He too rushed in one direction. In his direction, he found the pretty baby crying beneath a bush.

The baby, just abandoned by her mother, was safe as the birds were cooing and watching it nearby with care and concern for it. The songs of birds went on like the mother's lullaby appeasing and pleasing it. The sight of it was spectacular for Dakshinya Deekshith. He happily took it into his arms and fondled it.

The baby's looks were fascinating and charming. He felt that he was lucky to have found the baby for him to adopt. He therefore felt full delight instead of feeling deep sorry for its abandoned plight. He said to himself while the birds were cooing joyfully all around,

"I'm fortunate to find the baby and the woman who abandoned it is unfortunate."

"The poor lady left it today in the early morning for me to sight it."

"Perhaps the mother had seen us taking rounds here, expecting us to see it surely."

"I love it most to adopt it as my daughter."

"How charming you are, my dear baby!"

"Its parents must be charming with their winsome qualities."

Dakshinya Deekshith took the baby into his arms and went home happily to show it to his wife, Annapurna, as he wanted to adopt her as his daughter. He showed the baby to his wife earnestly. She was happy at its sight and took it into her lap.

It was the first baby abandoned by its mother for Dakshinya Deekshith to find and adopt it as his daughter in the orphanage. He decided to treat her as their third daughter as they had had two daughters Suma and Sushma born that year. His wife was immensely happy for adopting the baby. He felt that it would be the happiest occasion in the history of the orphanage.

On an auspicious occasion, Dakshinya Deekshith named the baby Shasima. All called her Shasima fondly. She grew to please the parents and the inmates of the ashram by her pretty looks and tender smiles.

When Dakshinya Deekshith found the first abandoned baby, he brought up it with all love and affection. His wife, Annapurna looked after it, fondling it very often.

Dakshinya Deekshith found the baby very charming like the moonlight and so he named the baby Shasima after the moonlight, as she looked very bright like the full moon. The baby grew to be a brilliant child.

The parents observed Shasima to have special likings. As a child, she loved the objects of nature. She enjoyed sights, sounds, and scents in nature. All joy glowed in her cheeks at the sight of nature. Nature with its sights, smells and thrills of flowers was the inexhaustible source for Shasima's pleasure.

Shasima looked at the sunrise amid the barred clouds. She felt delighted by the lovely young rays amid the flight of birds in the sky. It was the most enchanting scene for her smiles. She loved the clouds that moved in grace in sky.

As a master, Dakshinya Deekshith had established the orphanage in the lap of beautiful nature last year and managed it successfully. He had a lot of magnanimity, as he was a man of humanity to say,

"I don't leave any body as an orphan."

"It's my daughter or son to grow in my orphanage."

Dakshinya Deekshith evinced a keen concern for the babies abandoned by mothers. His wife, Annapurna co-operated with him in all respects and became the main supporter to him in reaching his cherished goal of bringing up abandoned babies.

* * *

9

When Laila left Vishesh to his fate, Vishesh fell into the sea of tears. He felt away from the hive of love like the fish out of water. The separation of lovers was like the cyclone to separate the pair of birds in life. His life fell not only into the sea of tears but also entangled in the realms of confusion. He felt extremely sorry for his abruptly broken love. His life appeared a desert without any oasis in his state of confusion tinged with the peeps of sorrow. He felt his life at the crossroads.

Vishesh did not know where to go and what to do then. He was able to see only the road of success to remind him of his father Parameshwar. Immediately he felt relief and solace in recalling him in times of adversity.

Vishesh recalled his father's advice not to join politics and realized that it was his utter fault. He memorized and revised the principles he learnt at Shantinikethan. He thought that his father, Parameshwar, was right in all respects. He went back to his father and sought his apology. He was on his mission to look after the agriculture land, and he was on the welfare of his father's workers. He was happy to be a farmer as the backbone of the nation than to be a leader to encourage deception, corruption, and exploitation against the wishes of the people. In fact, he felt that the farmer was the leader.

Parameshwar educated Vishesh at Shantinkethan with all expectations that he would learn all ideals. His knowledge of all ideals would be useful in enlightening his workers and their children after the completion of his studies at the institute.

As per his father's expectations, Vishesh taught his workers and their children to be educated and proved himself to be a farmer-teacher and leader-teacher rather than to be a political leader in the modern scenario. In recalling his father's words, various feelings were flooding in his mind,

"The farmer is the king of the people for their well-being. The farmer in the welfare of the people is a leader indeed."

"When the leader teaches his people how to earn their livelihood sincerely, he is the leader par excellence."

"The leader is supposed to lead the people to reach their goal but not to offer them the sops and freebies and turn them dependent on him."

"The people should live independent lives. They should contribute something to their nation on their part also."

"The farmer is above all others. He's the life-giver. He's the bread-giver. He's also the leader. He is the farmer-leader. He is the teacher-leader to the core."

Vishesh was a leader indeed to lead the people to earn their livelihood and lead the lives on their own rather than to receive offers, the temporary benefits: alms or sops in the reach of one's political goal to meet their needs and fulfill their vote-bank dreams for the time being.

Vishesh grew to be a model farmer on the lines of his father's advice. He was with his father Parameshwar and mother Prasanna. They wanted to perform his marriage with a match of his choice. He expressed what happened to him in the recent past,

"I loved Laila and married her. We led married life for some time. It was our marriage to be the marriage lifelong. I feel that she is my wife even today."

"We weren't aware of your marriage," said Vishesh's parents.

"She wanted our marriage to be a secret," said Vishesh.

"Why did you feel it a secret?"

"Her advice to keep it a secret sounded peculiar even to me. I felt sorry for that," said Vishesh in the choked voice.

"She doesn't know the bond of love and the nexus of marriage," said the parents.

"She bid goodbye to our love and our marriage for some time, leaving me."

"Where did she go?"

"She left for her place to lead her life in full freedom," said Vishesh.

"Why didn't you prevent her from going to her place?"

"By the time I woke up one day she wasn't there by my side. She left for her place to live happily.

"You would have searched for her," said his mother, Prasanna.

"She wrote a letter and left it in bed, before leaving for her place. She stated in her letter that I shouldn't search for her. She didn't appear again."

"What was the reason behind her leaving?"

"She wanted to live in full freedom in her region."

"You were in love with her. She met and married you secretly. You lived happily for some time, and she left you when she wanted. It appears some plot was there behind all happenings to you."

"Yes, some plot was there behind the screen. Indrasen, the son of Suryasen plotted to send her, the beautiful woman to me asking her to infatuate me by her beauty as he didn't want me to join politics and so he plotted to distract my attention in contesting the election and filing nomination papers. I didn't contest the election to be in the fray and so I wasn't his political rival. He was successful in that secret plot."

"Forget what happened. Now you can go for your second marriage."

"No... my dear parents...I remain as Laila's husband in life. Laila is my wife... Laila is my life partner...Laila is my beloved for my love forever... I serve my workers of agriculture as per your wish."

"When you aren't at fault, you can marry and have the second wife of your choice. She'll be your life-partner to help you in all respects. It's our wish,' said the parents earnestly.

"Laila is in my heart though she came to me on her mission...She descended on me as per the secret plan of Indrasen. She left for her region after her mission had taken a clear-cut shape."

"Isn't it deception?"

"I don't feel it deception as she offered me all her pleasures in the period that are to be offered in her whole life. Our union of course is short as per timescale, but it is lifelong as per the sweet memories she left with me."

"There're such women to love honestly and deeply."

"I've only one wife though she lives with me or elsewhere. She always abides in my heart, delighting me by her sweet memories."

"You don't understand what I say to you."

"Correct, I don't understand what you say."

"How long do you live all alone?"

"By serving our farmers as a farmer..."

"We wish that you should have your life-companion."

"Laila is my life-companion. I keep the doors of my heart open for her to re-enter into my life."

The parents: Parameshwar and Prasanna were unhappy for their son Vishesh's firm decision not to go for the second marriage.

The parents felt hurt for the plot of Indrasen in spoiling the life of their only son, Vishesh.

The parents had innumerable feelings in their hearts as they faced many clashes between what they aimed and what took place in their lives:

"Man is the biggest enemy of man. Indrasen as a fellow being wanted to spoil the life of my son, Vishesh. Let my son not join politics. I wish that my son would serve our workers, farmers since we are in their welfare forever," said Parameshwar in the choked voice.

"We haven't done any harm to any man...not even to our enemies. No God...no goddess evaded the plot of Indrasen against my son," said his mother, Prasanna.

"Man with principles becomes a scapegoat. My innocent son has become a victim to someone's plot," said Parameshwar with a deep feeling.

"Vishesh deserves a happy life as per his virtues but suffers for no reason," said mother, Prasanna while tears were rolling down her cheeks.

Vishesh accepted life as it happened to him. He was ready to face life without any grumbling or complaint or comment. His ship of life voyaged in the river of life in that manner, as he was an ideal lover for his beloved.

Vishesh did a lot of service to his workers, the world of farmers. His father supported him in all respects. He had a plan of his own to develop as a farmer in the welfare of his farmers.

In the service of the farmers at Karshak Kshetra, Vishesh rendered invaluable service and won name and fame.

One day, Vishesh discussed the plan with his father. His father agreed whole-heartedly to his plan of his son.

The next morning, Vishesh called his workers. All assembled in the field in the lap of nature in the presence of his father.

"You, dear workers, you've been working for us since you were children. You don't know any evil plan to tarnish our image. We've a lot of yielding because of your committed labor and relentless work. Since you're here with us as sincere workers, we've a lot of crop production and progress. Many other farmers are very jealous of us...

It has been my father's wish to make you all the partners of our agriculture. The land we've now is the land of all of us. Hereafter, we all, including my father Parameshwar and me as his son Vishesh, are the partners of our agriculture...We're the owners of the land. We all work together and enjoy the fruits equally...

We're the members of a big family. The land is ours. Nobody, including my father Parameshwar and me Vishesh has the right to sell or mortgage it or a bit of it for any purpose. We can meet all the needs with the income of our agriculture. We cultivate the land and enjoy the fruits of cultivation all together...

Our income is for the welfare of all the partners of our agriculture.

Our health is very important. We have health insurance for that we should invest ten percent of our income for it...

For the education of the children in Karshaka Kshetra, ten percent of income is spent from the income of our agriculture. All the children get the facility. All have education and a particular parent need not think of the education of his children. It is the outlook of all the farmers all together...

For functions, ten percent of the income of agriculture is allocated. They celebrate the functions in a befitting manner all together...

There is a major allocation for the construction or repairs of projects, buildings, tanks and so on. Twenty percent of agricultural income is for that...

Fifty percent of income is for the welfare members of the family of Karshaka Kshetra. For food, clothing, housing and others, the rest of the income is ready. This is the financial plan of Karshak Kshetra expenditure." said Vishesh in the welfare of Karshaka Kshetra.

"No, we work for you. We're for your development. You treated us as members of your family. You never treated us as your slaves unlike many landowners elsewhere. We feel credit to work in your land. We aren't the owners of this land," said the workers in one voice.

"All the temples of sculpture-marvel were not built by kings. They were bruit by workers and sculptors...

Leaders didn't build all multipurpose projects of a nation. The workers and engineers built them...

See, our land is cultivated by us all. We've all agricultural income because of you all... your toil...your sweat...Hereafter, we as the partners enjoy the income of our agriculture.

Agree with me. We'll be model farmers. We set a model to the farmers in all places. We all take credit for that," said Vishesh with all feelings of deep commitment.

"We never felt that we were workers. We felt that it was our land and we worked with the sense of commitment. We'll continue to do so in the future," said the workers.

"Let's set a model to the world. All should follow my model plan. We all get name and fame as the partners of agriculture. It's good for all of us," said Parameshwar, and Vishesh nodded his head.

Then the workers felt convinced and agreed to the plan of Vishesh. They gave a big round of applause.

"We feel our responsibility enriched to the highest degree to work with the utmost commitment. We welcome it to bring name and fame to our great farmers, Parameshwar and Vishesh," said all, clapping all together.

"Thank you for accepting my proposal in the welfare of all farmers in the partnership of ideal agriculture," said Prameshwar and Vishshesh.

"You're great. We wish that you would be the sense of inspiration for all zamindars or big landowners in the nation and the world," said the workers in joys.

They all celebrated the occasion in a befitting manner in boundless joys.

All the newspapers appreciated the proposal of Parameshwar and Vishesh like anything. They expected that it would be inspiration for all people. The newspapers applauded them as "Rithu Bandhu"to mean “the dear of the farmer” in the real sense. Some papers said it was a "great sacrifice" on their part. Some papers appreciated them as the "Real Leaders". They made this "great decision in the days of land-grabbers". Some appreciated them to be "the followers of Vinoba". Some praised them for "a novel and model" plan and were happy to find such people in this society.

The Prime Minister sent a message stating that Parameshwar and Vishesh were model famers to leave inspiration to big landowners. All the other important leaders appreciated them on the lines of the Prime Minister's appreciation.

The land's grain yielding was more and more year after year in the committed work of partners in agriculture. It came to the notice of many farmers. The Government recognized the plan of Parameshwar and Vishesh as the best. They felicitated them as "Rithu Bandhu", the dear of the farmer at the centre.

... ...

Mr. Vijay after his victory in the election, he met Vishesh in his house. He sought his hearty wishes and cooperation,

"Mr Vishesh, I've first come to you to seek your blessings," said Vijay,

"Wow, you've won the election, Congratulations to you my dear friend," said Vishesh happily.

"It's your victory...If you had contested you would have been the winning position," said Vijay.

"I hoped to win and wanted to contest the elections. Meanwhile a storm raged in my life," said Vishesh with some feelings.

"I know the storm. I know the man who raged the storm,"

"One needn't be a political leader to serve the people. He can serve the people as a farmer, as a teacher or as man of humanity. I started to believe in that," said Vishesh.

"Exactly...when man isn't man to be human, he's a brute, brutal like wild animals...worse than wild animals. In fact, wild animals do not harm their own species. It's man alone who can deceive his fellow man. He as man proves that man is the biggest enemy of man," said Vijay.

'That's the true study of man's race," said Vishesh.

"The political field needs a person like you, I welcome you to this as it's the platform for a leader to use his powers, machinery and all," said Vijay.

"Of course, you're right in your aspect. That's what I felt when I decided to contest the election," said Vishesh.

"Your victory was sure as the people posted their messages in the social media. I heard and read all those in the newspapers," said Vijay.

"Though you win. or I win, it makes no difference. Our people must be happy with your selfless service and model governance. They must be free from humiliation and harm. or loss to anyone. That's the prime and primary goal of a leader," said Vishesh.

"Yes...That's the right approach of a leader elected. I like your policies in making workers as partners. I congratulate you on making them partners. I adore your sacrifice. The central government rightly felicitated you on the policy with the title, Raithu Bandhu. You deserve it cent percent. You're a model farmer...You're a farmer-leader and farmer-teacher. You've made true the saying, "The famer is the leader." You're the leader indeed," said Vijay.

"Thank you for your keen concern for human welfare," said Vishesh

"You're the chief inspiration for me in many aspects," said Vijay.

"Thank you... Every man should be an inspiration for at least one man for a good purpose. Then his life is worthwhile as a human being," said Vishesh.

"I seek your wishes and blessings for my success," said Vijay.

"All the best...You prove to be successful leader," said Vishesh.

"I thank you very much," said Vijay bowing the head.

* * *

10

Dakshinya Deekshith brought up the abandoned babies as his children. He adopted thirteen female babies including the first adopted child, Shasima and three male babies as his children. He brought up them with loving care and concern. In the coming years, the number of children in the orphanage grew up.

As the lover of education, Dakshinya Deekshith gave all the children all good education. The children had all round development in their effective teaching. They grew to be very brilliant with extraordinary talents.

Annapurna, his wife also treated all adopted children as her biological children: Suma and Sushma who were twins born that year. They learnt all to be well mannered and good-natured to be worthy children of worthy parents to bring them name and fame.

Dakshinya Deekshith wanted to perform their marriages after giving them all ideal education by his excellent instruction.

The orphanage was famous for its excellent education. Dakshinya Deekshith was such a teacher to impart all worthy things to shape them into worthy students with their multi talents. They were keen on participating in their curricular, co-curricular and extra-curricular activities.

....

Shasima proved to be a prodigy. She had all exceptional qualities. Dakshinya Deekshith noticed her liking nature and her positive response at the sights and sounds of nature. When there were dark-blue clouds, she waited eagerly to watch the lightning amidst them. As soon as she looked at it, her face glowed in excessive joy. She sometimes appeared to speak to the clouds,

"You are mothers pregnant with the wombs of rain to shower life for lives."

The rainbow with its seven pretty colors was a special attraction for her. She felt it was a bridge for her to ascend. It was a spectacular sight to fill her face with glows and her heart with joy.

Shasima loved to play with her sisters when it drizzled amidst the rays and the rainbow was to glow in seven pretty colors. It was an enjoyable and memorable sport for them as children.

The objects of nature were Shasima's dolls. She as a child developed love for the singing birds, humming bees, fluttering butterflies, comforting breezes, dreaming buds, blooming blossoms, cascading waterfalls, loving fawms, dancing peacocks, and others for their lovely sounds and loving relations in the lap of nature.

Shasima lent her ears to the songs of sonorities. Whenever she heard the songs of cuckoos, she felt thrilled. She danced to their tunes sometimes. She tried to imitate their sounds. She also inspired the songbirds to sing by singing like them.

Shasima loved birds in variety with their melodious songs and their beautiful feathers in varied hues. When she called parrots, they descended to perch on her shoulders. She imitated the coos of singing birds. She inspired parrots to coo some words and expressions like "Rama", "Shasima" and "How do you do?"

At ease, the parrots cooed, "Rama", "Shasima," How do you do?"

Shasima learnt the fact that nature is the treasure of treasures. She called the birds by their sounds in imitation. She praised the cuckoos,

"None...no bird can sing like you, my dear sweet-throated birds. I'm proud of learning your voice even a little."

Shasima liked small-winged creatures. Bees hovered around her, humming. She loved their humming. She moved in grace taking rhythmic steps in joy. She was in full joy in the breeze blowing gently. She was happy amid the melodious sounds in nature in all seasons especially in the spring season.

Humble bees, butterflies and bees knew why they loved her. The sight of humble bees, butterflies and bees was a pleasant sight for her. She loved their flight, especially that of butterflies around plants. She tried to catch them in her palms. They kissed her palms and flew to the beautiful flowers.

Shasima played happily in pastures in sheens and verdant plants with beautiful blooms. When she looked at the flowers, her face glowed with dimples in her rosy cheeks in immense joy.

Shasima touched her cheeks to the bloomed flowers. Their softness thrilled her cheeks. The flowers scintillated in her eyes with their hues as the beauty in a rich variety. The pretty sights hosted the delightful feast to her eyes. She was the lover of flowers. She was the flower of flowers because the butterflies as the lovers of flowers loved her very much. Humble bees hovered round her as they thought that she was the flower of flowers. She often felt surprised at their boon of blooms all around. She now and then spoke to flowers silently, appreciating their unrivalled charms,

"O flowers, when the earth is without you, the earth is without mirth. You fill the earth with all charms and scents. Without you, my dears, the earth is like a desert without any oasis. Nature celebrates a fest with flowers of beauties in varieties for my gaieties."

Shasima felt thrilled at their sight, appealing to them to grant their grace and elegance to her,

"O flowers, you have beauty in plenty. You can lend me your beauty to me to look like you."

In reposnse to Shasima praises, the flowers seemed to reply in joy,

"You are as beautiful as we are in all respects. We sway on the twigs only, but you move everywhere so beautifully, dancing gracefully. You're the moving flower...the flower of flowers in the lap of nature."

Shasima smiled heartily at the response of her dear flowers, kissing them by her rosy lips in all love for them.

There were creepers with flowers in the surroundings of the orphanage. They were a special attraction for her. She loved the creepers and danced amid them. She bent her body like creepers in her dance and walk.

The inmates of the orphanage, mainly Shasima, did not allow the plucking of flowers, the smashing of creepers and the hunting of animals in the surroundings of the orphanage. Shasima and her sisters loved trees, treating them as friends and well-wishers to offer oxygen, fragrance, fruits, and cool shade.

Inhabitants of the orphanage planted the saplings where there was room for them to grow well and wide. They named the trees after gods and goddesses. They named some saplings after their names. They watered the plants to grow exuberantly.

They felt sorry when anybody harmed the plants and cut down trees. They watched them grow exuberantly, enjoying their sheens in greens. They thought that the trees would help them live: *vruksho rakshathi rakshitha*. They treated the trees around the orphanage as their mothers.

There was the lovely scene of beautiful waterfalls for the inmates of the orphanage. They loved the waterfalls very much. The beauty of waterfalls filled their hearts with joy. Shasima looked at the waterfalls getting engrossed into their falls in grace. She praised the waterfalls,

"Wow, I'd like to leap from the top to the bottom like you in my might."

There were hills, seen from the orphanage in north, east, south, and west directions as a special attraction from distance. The hill in the north appeared like a big dark-blue snake crawling without moving. The hill in the east looked like a big tree welcoming all to its cool shade. The hill in the south seemed to be grain heaped into a store by a farmer in one place. The hill in the west was like the bull with its lovely eyes to look lively, relaxing joyfully in the sitting position. The inmates of the orphanage enjoyed the sights of waterfall and forest nearby. It's joyful to look from the orphanage in all directions.

The orphanage and its surroundings were very beautiful to look at. Shasima and all others lived in gaieties amid the beauties and sonorities of nature.

Among all birds, Shasima loved peacocks and observed them calling her to them. She spread her hands gracefully in her dance posture in the way the peacocks unfurled their beautiful feathers. The sights of peacocks in dance were the source for her grace. She referred to the peacock, raising her eyebrows,

"I love your dance...your graceful dance... the unfurling of your feathers."

Of all, Shasima loved young animals like fawns in nature. She played with them all around the orphanage. She enjoyed the company of all lives in nature. She loved her pet fawn very much. It was actively participating in her play every time. She expressed her fondness for it,

"In your company, my happiness knows no bounds. Your love for me is boundless, my dear fawn."

"You're my friend. Without you, my life will not be very joyful."

"Next to you, I love you all others like cows, birds and other young animals."

Dakshinya Deekshith had cows. The cows grazed in the lands around the orphanage. He had a boy to look after them. To look at the orphanage and surroundings with the cattle was pleasant for the visitors. Shasima called cows and calves by their names. When she called the cow by its name, it rushed to her and she fondly touched it, saying,

"You aren't just an animal. You are mother, the mother of mothers."

The inmates especially Shasima fondled calves very much. She played with them with all love for them. She treated them as friends,

"Without the company of calves, my life would have lost something special. Luckily, I've your company. I enjoy playing with you as my playmates."

Dakshinya Deekshith built the orphanage in the place he wanted to build. He had cows graze in the surroundings of the orphanage. Wild animals came to graze in the surroundings of the orphanage.

Though the orphanage was on the outskirts of the forest, the wild animals did not do any harm to any life in the orphanage. They grazed around the orphanage and went inside the forest in the night.

The orphanage inmates loved fruit trees, flower plants, birds, and animals very much. Their days passed happily in the company of all those. Theirs was a beautiful garden on the outskirts of the forest with the sights of flora and fauna. They lived in the lap of nature. For them, nature was their mother. That is why they adored nature with glories as their goddess.

The inmates of the orphanage were not just nature lovers but also knowledge-seekers. Shasima and others were good at learning and obtained enormous knowledge. She stood first in the public examinations. She was competitive enough and others felt incapable of competing with her in any manner. All considered her an A-1 student.

Shasima and other students of the orphanage participated in debates, elocutions, group discussions, G.K. competitions and so on. They participated in all the competitions and received awards and accolades at all levels. Shasima was good at speaking fluently and mellifluously. Her speeches were very impressive to the listeners.

The orphanage was famous for its participation in games and sports. Shasima was good at playing chess. She won awards at the national levels. Others got prizes. They had all round development to be competitive enough in any field.

Shasima was the best award winner in the state and one of the best award winners in the nation. She received the certificate of appreciation and the awards as the best participant in the state.

Dakshinya Deekshit loved Shasima for her multisided and multifaceted talents. He liked her very much for her extraordinary qualities and charming manners. She was unrivalled in the orphanage.

Shasima was in the orphanage. She was very happy with the love of her parents, Dakshinya Deekshith and Annapurna. She did not know that they were not her real parents. She knew them as her parents. She was the most lovable girl to the parents in the orphanage.

Shasima was the fond daughter of Dakshinya Deekshith and Annapurna. She never knew that Vishesh and Laila were her biological parents.

Laila was leading her life in the region of freedom. Vishesh was at Karshak Kshetra working as one of the partners of agriculture. They were in two different goals.

* * *

11

Malli's parents approached Suryasen, seeking justice in their daughter's issue. It was of no use. He paid a deaf ear to the issue.

During the election campaign, Suryasen always said, "Let's wait for some time to take up the issue."

Suryasen deliberately postponed the issue until the elections were over. He thought that the raising of the issue would adversely affect his son's winning chances in the election.

Malli's parents had no way except to delay in raising the issue. Moreover, it took time to make Malli's issue known to all the people.

Elections were over. His son, Indrasen tasted the debacle despite his most possible efforts. He spent crores and supplied drinks but failed to win the elections.

Suryasen was inside. When Malli's parents sent a message to him, he came out with fury in his face.

The parents reminded Suryasen of Malli's issue. Her pregnancy was due to the deception of Indrasen. He molested Malli and had sex with her. As a result, she beame pregnant.

"Who said so?" said Suryasen.

"Everyone in your house, your relatives and the people said so. They know this issue even today," said the parents.

"Who is Malli? Who is Indrasen? There is a lot of difference between the two. He's a prominent leader...He's the son of a leader who represented the constituency for two decades. He's a big landowner. What is Malli? She's a servant-maid. You talk of their relation. How is it possible? Shutting your mouth, you can get lost," said Suryasen and went inside.

All the people who supported Malli's cause shouted at Suryasen not to go inside. He did not come out, caring not even a bit.

"You can send your son, Indrasen out," said the people.

The people waited for some time, expecting Indrasen to come out. He was the person for all to cast their votes for him.

Indrasen did not come out. The people were angry for the negative response. They shouted at Suryasen and Indrasen in their ways:

"Indrasen, come out...Come out with your father, Suryasen,"

"You're leaders. You're the leaders of the people. You're to lead the people but not to deceive...not to exploit them. Mind the truth."

"You can come out to speak to us. You as leaders can listen to my feelings."

"You're bound to grant us a chance to speak to you, our leaders."

They did not come out. They paid a deaf ear to the appeals and orders of the people. They felt angry with the leaders. They expressed their feelings of anger to the leaders inside,

"You are leaders...You aren't ordinary people..."

"Like demons, you played with the lives of the innocent people. O Indrasen, you pose to forget the issue of Malli... You can't forget how you spoilt the life of Malli."

"Indrasen, you can see your son as an infant in the lap of Malli. All know about your father, and you and the infant. All know everything, especially the infant, your son, growing in Mall's lap. He's very much present to ask you, 'Who's my father?' All know that you're responsible for the infant's birth and you're his father."

"Do you think that others don't know about your deception...your secret deception within the walls of your house?"

"You can't hide what you had done in your house, spoiling Malli's life, telling cock and bull stories. The innocent victim is suffering outside. Come and see her shedding tears."

"When your father performed your marriage secretly with a girl from the city nearby, didn't you remember Malli whom you spoilt to satiate your lust?"

"Are you a leader...? Are you a human being?"

"O Suryasen, we know all your concubines and your secret relations with them."

"Innocent Malli didn't know you have concubines. She wasn't aware of your son to be barbaric in nature. He spoilt her life."

"You are rogues... You are cheats...You must be ashamed of being born as men and having the human form."

"You're cow-faced tigers, goat-skinned wolves."

"Your true colors are hidden under guises. People, surely, see your real colors one day."

"You're dragged to the crossroads to face utter humiliation."

"We're going to the court of law. We're going to the government."

"All the people are ready to support us at any cost. You can't befool anyone hereafter. We know your real colors... your true hues... You are chameleons to change your colors and crocodiles to shed tears. We can't believe your words as you are cruel fellows to use any one to satiate your lust."

All the people immediately went to the newly elected leader Mr. Vijay. The newly elected leader came out with folded hands to listen to them attentively. They narrated all what happened to Malli,

"Malli was the servant-maid in the house of Suryasen and Indrasen."

"Indrasen was responsible for the birth of the infant growing in her lap."

"Suyasen performed his son, Indrasen's marriage with the daughter of a leader even after Malli was pregnant nearly a year ago."

"After her pregnancy, we approached the parents of Indrasen, but they didn't care for our appeals to them in their house."

"They simply said, 'Let's wait for some time to look into issue'."

"Before the date of elections, we went to him, he repeated, 'Let's wait for some time to look into the issue'."

"During the election campaign, he was to say, 'Let's wait for some time to look into the issue'."

"He repeated his tactful advice for us, 'Let's wait for some time.' We decided not to wait any more."

"Now we understand why he postponed the issue. We asked him barefacedly to do justice to Malli."

"Now, Suryasen has said today, 'There's no relation between his son, Indrasen and the servant-maid, Malli. It's impossible for him to have that relation as she belonged to the low caste', saying so he went inside."

Vijay listened to it very attentively. He said," I'll look into the issue and see that justice is done to her."

The people went home and discussed their plan to go the court of law as per the necessity. They made up their mind to achieve their goal in support of Vijay.

They gave assurance to support Malli until justice was done to her.

Mr Vijay advised them to approach the court of law. He advised them to consult Chakravarthi, one of the most famous advocates. They consulted him as per his suggestions. He practiced as a lawyer and had experience in dealing with such cases. He spoke to the advocate to see the justice done to her.

Mr. Vijay, the elected leader advised the lawyer the way to win the case. He hinted certain things for early judgment only by means of DNA reports.

Chakravarthi, the lawyer of Malli sent a notice to Indrasen to prove that he was not guilty in Malli's case. Indrasen and his father approached a lawyer related to them. He was Jitendrasen, known for his manipulation to win the case. When they requested to save him, he demanded a lot of money. He said to them,

"You've to manage to get a wrong DNA report from the doctor concerned."

"Yes, we can manage," said Indrasen.

When DNA reports were ready, the court of law examined them. As per the reports of Malli's son and Indrasen, it stated a difference.

Indrasen managed to get a wrong DNA report. Jitendrasen challenged the report in the court of law that Malli's son was not the son of Indrasen.

In the presence of the Judge, Chakravarthi made an appeal to the doctors concerned. He wanted the DNA reports. At that time Indrasen was not able to manage as the doctors felt that they would be at risk. That time they gave correct reports.

The DNA proved that Malli's son was the son of Indrsen. The court was ready to award a punishment to Indrasen. Then Indrasen had no other alternative and so they were to honor the judgment pronounced in the court of law. He accepted that Malli's son was his son.

Indrasen married another woman after Malli had got pregnancy by him.

As their living together was true, Malli would be the legal wife. She's every right as Indrasen's wife.

The woman who married Indrasen later had no right to claim to be his wife. As per law, the marriage would stand invalid.

The tragedy was that Indrasen's second wife already had pregnancy by him.

Malli would be his legal wife. She would have every right in the properties of Suryasen as the wife of his son, Indrasen.

… … … … …

After some time, the court pronounced a judgment that Malli would get half of Suryasen's house and fifty percent of his landed properties, and all these halves belong to her legally.

Malli occupied half of Suryasen's house. She started to live in the house. She started to cultivate half of Suryasen's land.

The judgment gave a deathblow to Suryasem and Indrasen. They stayed indoors in utter shame and humiliation.

.

Malli was a woman of innocent nature and human concern for the people especially the poor and the landless. She read and heard about Vishesh's great policy of making his workers the partners and owners of his father Parameshwar's land. The policy impressed her a lot.

One fine morning Malli declared that she would follow the policy of Vishesh. She made all the workers as the partners of her land that she got as the wife of Indrasen. She became a true follower of Vishesh.

The people noticed a lot of differences between the character and the nature of Vishesh and those of Indrasen and Suryasen though they were close relatives. Vishesh was the son of Suryasen's sister Prasanna and Indrasen was the son of Suryasen. They were diametrically opposed to each other in all respects.

The people commented on Suryasen's family especially Indrasen while complimenting Vishesh on his remarkable works,

"A man is known by his character. Laila is infatuated with Vishesh. She left him for no fault of his. He didn't go for his second marriage. He further treated Laila as his wife. "

"Indrasen raped Malli, his servant-maid. As a result, he was blessed with a son, but He doesn't accept her as his wife."

"Vishesh is a model farmer. He treated his workers as members of his family. He made his workers as the partners of his agriculture. How great he is!"

"Indrasen is a dictator like his father, Suryasen in all respects. He treats his workers as slaves. He treated Malli, his servant-maid, as a slave and spoilt her. He doesn't treat her as his wife. How cunning he's!"

The people had a very bad opinion about Suryasen and Indrasen. Malli who had looked after them with all care and concern as their servant-maid, treated her as cruel fellows.

Malli performed the naming ceremony and named her infant son, Harsha in the presence of Mr. Vijay, the newly elected leader, the Legislator of the constituency and many people of the village but not in the presence of the members of Suryasen's family.

The little boy, Harshith, was very handsome. He inherited his mother Malli's features. He grew to be a handsome boy.

Malli was successful in getting the half share of Suryasen's properties with the help of Mr Vijay and the people of her village.

As a girl of a poor family, Malli had ideals of her own. She had the qualities of head and heart. She treated her rapist as her husband though he did not consider her his wife. She knew family vaues and virtues. She at the same time wished for the welfare of all people. She did not like any harm done to anyone by her. She did not love the triple evils: humiliation, harm and loss meted out to anyone. Her heart shared the feelings of the victims of triple evils.

Malli was therefore against the wicked fellows: Suryasen and Indrasen.

As a girl born in a poor family, Malli read about great leaders like Ashoka, Mahatma Gandhi, and Vinoba. She learnt about Vishesh and his ideals. She wished that all big landowners would follow Vishesh. They should appreciate his policy of making the workers partners of his agriculture.

For Malli, Vishesh was her role model and followed his principles to the core. She was unhappy with the bad behavior and dictatorial attitude of Indrasen.

Malli was under the influence of the plans of Vishesh. She followed his footsteps to make her workers as the partners in agriculture.

The newspapers appreciated Malli for her sacrifice. Though she started her life a servant-maid, she had good heart for her great sacrifice.

* * *

12

Time rolled on its wheels non-stop and found the young ones like Suma, Sushma and Shasima in the orphanage grow and shine in youthful exuberance and charmful excellence. Their growth into damsels reminded Dakshinya Deekshith of his prime and primary responsibility of performing the marriages of three daughters: Suma, Sushma and Shasima.

As a father, Dakshinya Deekshith gave them a very good education. He taught them how to respect their in-laws, life-partners, and all elders. They learnt to be well mannered and well cultured as their father taught them with care and concern.

The orphanage witnessed a lot of progress in all respects under the auspices of Dakshinya Deekshith. Shasima, her sisters and brothers were good at curricular, co-curricular and extra-curricular activities along with household chores.

Dakshinya Deekshith used to go to other orphanages and residential schools and colleges and delivered extension lectures on the importance of residential education for imparting wisdom-oriented-knowledge, culture-oriented-awareness in shaping the students into ideal citizens.

This time Dakshinya Deekshith went to Kolkata on invitation to deliver his lecture on the residential system of education and its unrivalled merits.

Dakshinya Deekshith assigned the responsibility on the shoulders of Shasima to look after the orphanage and its activities in his absence. He was away to lecture in other orphanages and institutes whenever he had invitation as he encouraged the residential system of education. He had firm belief that it could establish values and virtues, morals, and ethics and so on. All respected him for his exemplary and exceptional stature.

During Dakshinya Deekshith's absence, Shasima looked after the orphanage. She received and entertained the guests as per her hospitality. The activities went well in her supervision. She looked after the cows.

Shasima grew to be a very charming damsel. She was in her youth. She played games and sports with other girls in the surroundings of the orphanage. All loved to play with her, as she was their loveliest playmate. She was at the same time the most lovable girl in the orphanage. They enjoyed the sight of birds and animals while watching the cascades of waterfalls near the orphanage.

As part of her play, Shasima was with her a fawn, fondling it. When she was the sheens of greens, the butterflies were hovering round her for she was like a beautiful flower. She was happy to find the butterflies as its friends hovering round her.

In his absence, a wonder took place for which Dakshinya Deekshith was happy after his coming to the orphanage from Kolkata.

Shasima was in a delightful mood. Then a group of boys came from the city to visit the waterfalls near the orphanage. In the group, there was a charming boy seen amongst the boys. He was none but the only son of the richest and noblest farmer-teacher and farmer-leader of the nation. He had just come out, completing his higher studies at a nation-famous residential institute, Takshasila. It was named after the the famous university in the past. He was very handsome with winsome qualities. He was brilliant and his lovely face reflected his brilliance. His dimples glowed when he smiled. Amid all verdant scenes, he appeared like the rising sun on the eastern horizon.

The beauty of Shasima fell in the sight of the most handsome boy like the lightning amid the dark clouds. It was her unsurpassed beauty to draw his attention to her while she was playing with fawns near the pilgrimage and her sisters were wateing the plants. He came to close to them as he wanted to speak to them and opened his conversation,

"The place is calm and quiet near the waterfall that we came to visit. Here is a hermitage worth visiting in the lap of nature."

Shasima minded her play with fawns, bowing her head. Suma, her sister responded to his words:

"This is the hermitage of Dakshinya Deekshith to look after orphons in the lap of nature."

"Who are you to visit this hermitage?" said Sushma.

"I have come here to visit the beautiful waterfalls nearby," said he, looking at Shasima who excelled the other two in charms while she was listening to him in silence, observing him.

Suma and Sushma observed him observing Shasima with a keen attention to the admiration of Shasima.

The fawn with which Shasima was playing galloped towards him. She was to follow it at her pace in grace. She was moving like a beautiful flower swaying on the stem in breeze. Butterflies were howering round her face thinking that it was a multipetalled bloomed flower.

When Shasima came closer to him, he was so intent on opening an intimate conversation with her smiling,

"You are an angel on this earth... Where are youn from?

“I am a daughter of Dakshinya Deekshith from the orphanage, seen from here," said Shasima while bowing her head in shyness.

"O! You're a daughter of Dakshinya Deekshith about whom I heard a lot. I devoutly adore your father. He's more adorable than my father. It may not be befitting if I see him now. I specially come and see him the next time...This is a worthseeing place. I've met you for the first time here. I'm immensely glad to see you in the sacred place."

Shasima smiled heartily, expressing her thanks for his utmost gladness to see her.

"I am bound tell you again that you look like an angel descending from heaven and a mermaid from the ocean-treasure. I'm bound by your charms... I can't turn my glances from you to my dear fawns... You look at me in pleasure. Am I so handsome to attract you?"

"You can ask my eyes for the answer,"said Shasima.

"Your eyes say that you're able to reply in the sweet flow of your words, but you are in hesitation," said he.

"I don't have words. My words can't dare to describe your charms...You've dared to enter the premises of our orphanage... May I know your name?" said Shasima.

"I've already wished that you would ask my name. I'm very happy that you've asked my name. I'm Prashanth, the only son of the richest and noblest farmer-leader and farmer-teacher of the nation. In the nation, he's famous as a farmer like your father, Dakshinya Deekshith in the hermitage to look after orphons."

"When I asked your name, I too felt delighted to tell my name... I'm Shasima."

"The prettiest name for the prettiest damsel in the lap of beautiful nature..."

"Your father perhaps named you Shasima as you are like the moon and charms are like moonbeams…How beautiful you are!"

"You're like the morning sun with your radiant rays to brighten the face of nature with glows and charms."

"See, the buds have bloomed all around in the surroundings of the pilgrimage. Do you guess the reason?"

"I can't guess the reason."

"Your arrival marks the arrival of spring in the orphanage."

"Perhaps it's because of our union."

"Hear, the cuckoos are singing melodiously."

"They love our union."

"You're right."

While Prashanth and Shasima were in their intimate conversation, opening their sweethearts. Their dimpled cheeks shone to attract each other. Her sisters, Suma and Sushma watched her speaking to him in utmost attention.

Prashanth somehow learnt that Shasima was the daughter of Vishesh and Laila.

In the lovely impressions of Shasima, Prashanth was not able to leave her from their meeting place, as she was the heaven descend to please him by her presence. So was the case with Shasima in her deep impressions of him. They fell in love at first sight not to leave each other so easily.

"Shasima, it's the time for us to offer prayers to the goddess, Sri Vani."

Shasima joined her sisters when they called her as it was time for their prayers at the orphanage. As part of the orphanage activities, they sang hymns as their prayers, offered to the Goddess Sri Vani on its precincts.

It became inevitable for Prashanth to leave Shasima to join his friends. They all came to see the beautiful waterfalls in the forest near the orphanage of Dakshinya Deekshith. They went on their return journey to go home after their happy visit to the cascades of the waterfalls.

The inmates of the orphanage were in their prayers. They sang the hymns with all devotion, as they were to join the prayer every day as part of orphanage activities.

After the prayer, Suma and Sushma met Shasima met for their conversation in an intimate manner:

"This is the happiest day not only for three of us but also for all the other inmates of the orphanage," said Suma with all smiles.

"Yes, this's the happiest day. There's no doubt it," said Sushma in full joy.

"Today, our sister, Shasima had a lovely chat with the handsome boy we've seen so far," said Suma.

"What did he speak to you, Shasima?" said Sushma.

"Didn't you hear our conversation?" said Shasima.

"No, we were at a distance. We saw you speaking to each other from a little distance. We didn't hear your interesting conversation. We couldn't listen to your intimate conversation amid the songs of cuckoos," said Sushma.

"It's a surprise," said Shasima.

"It was something special," said Suma.

"The boy and you smiled to each other heartily," said Sushma.

"You were in full smiles," said Suma.

"Your and his faces glowed in boundless joy," said Sushma.

"You like the boy very much, don't you," said Suma.

"Definitely, you like the boy very much," said Sushma.

Throughout the conversation, Shasima kept smiling to the talk of her sisters. She understood that they were making fun of her speaking to the boy with all smiles. She had no answer except to smile all the time during the conversation.

Suma and Sushma were very happy as their sister Shasima found a boy of her choice for their conversation.

"Definitely, he'll come to see you tomorrow," said Sushma.

"I'm a hundred percent sure of his seeing you tomorrow, my dear sister, Sushima," said Suma.

Shasima kept on smiling for the pleasant fun made by her sisters: Suma and Sushma.

13

Prashanth again appeared on the scene the following day where he saw Shasima for the first time the previous day as per the expectations of her sisters. By the time he reached there, Shasima was playing with the fawns, but her attention was in distraction. He hid behind a beautiful tree and watched her secretly. The humble bees came in the direction, diverting her attention to him. Later he approached her gracefully.

Shasima appeared to be welcoming him and it was evident in her shyness and eagerness. Her cheeks shone more beautifully when she blushed, and her dimples glittered to enrich her unrivalled charms. He opened a delightful conversation with her with all smiles,

"How do you do?" said Prashanth with a smile.

"How do you do?" said Shasima with a glow on her face.

"How I was yesterday was better than how I was the day before yesterday. You can see how I'm today. I needn't tell you... I'm so because of...," said Prashanth.

"You say 'I'm happy because of you'."

"Exactly the words I wanted speak, flowed mellifluously from your mouth...," said Prashanth when his face glowed with dimples.

"The sun looks at us cheerfully," said Shasima.

"The sun shines brighter today at the sight of us together than before," said Prashanth in joy.

"Perhaps, the sun loves us to be together...our union... like the communion of two sacred rivers to flow a river to merge with the ocean-treasure," said Shasima.

"My heart was open yesterday for its expression. Today it's ready to express with your permission," said Prashanth, looking into her smiling eyes.

Shasima bowed her face, smiling. Prashanth understood her love for him. He opened his conversation,

"Shasima, I love you…I love you, Shasima. I welcome you to my life as my life partner," said Prashanth, stretching his arms wide to welcome her. She rushed to join him in no time.

"I too love you," said Shasima.

They hugged each other. They were in a warm embrace, knowing no fleet of time. Just then, her dear fawn came to Shasima, caressing her by its head, as it wanted her to fondle it. Its touch made them come to the world of reality from the heaven of embrace. She felt the coming of fawn unwelcome at that most joyous moment though the fawn was very lovely and lovable for her.

"As a token of my love towards you and our love marriage at the auspicious time, this is the ring. It's special for the inscription on it with my name and your name... Yesterday I met my nation-famous goldsmith asking him to prepare the ring in a special way the last night... I described you and your features to him. He was able to engrave you realistically in the ring as I described you...

Today I collected the ring from the goldsmith just before I started to come here...While coming here I looked at it and was very happy...

In making the ring, the goldsmith used his art and my heart in it to be the worthy ornament on your part to remind you of me...our marriage...

When you look at the ring, you and I, our miniatures, come to the centre from its corners and garland each other. Through the glass, we can see us garlanding each other. It is the ring to remind us of our love marriage. You can see the scene in it...

Wear it forever as the sign of our love marriage."

Shasima looked at it. She and Prashanth, their miniatures, came and garlanded each other. She was very immensely happy at the scene seen in the ring.

"Let's have a snap with us together," said Prashanth so, he put the ring on her finger with glittering smiles while taking a snap of theirs as the wedding couple: the bride and the bridegroom in their love marriage.

"Why don't you have a snap of us in your mobile? I don't use any mobile. When my heart calls you, you recall me forthwith. You dream of me. You fly to me on your wings," said Shasima.

"Exactly...Exactly...," said Prashanth in all joy.

Shasima's happiness knew no bounds. Their love marriage was the fest in the lap of nature, amidst her twin sisters and her dear fawn. The sun in the sky blessed them heartily. Every flower and leaf shared the wedding joys on the occasion by swaying gracefully in breeze.

They enjoyed the life of their youthful exuberance in the verdant nature. Nature witnessed them enjoying their married life.

Prashanth went back, promising Shasima to come the following day.

Shasima along with her sisters went to the orphanage happily. That day there was a cheerful atmosphere on the eve of the auspicious occasion of her love-marriage.

In the absence of Dakshinya Deekshith, Shasima entertained the inmates and the guests. The guests enjoyed her hospitality.

After prayers and all programmes, they had the habit of watching TV programmes on the doordarshan channel. That day, it telecast Dakshinya Deekshith's interview with the students at a residential institute in Kolkata. He gave answers to the questions of students in the TV programme:

"Dear masters and my students, I would like to present my views on the importance of the Residential system of education. It marks a clear-cut difference from other systems of education. For education it's only the system of education to set a model."

Like that, Dakshinya Deeksith spoke on the Residential system of education for an hour, welcoming questions from the students.

"Sir, why do you prefer the residential system of education?"

"The residential system of education is preferable because it's free from outside influences like sounds of pollution, strikes, and so on."

"What's the specialty of the residential system of education?"

"Peaceful atmosphere prevails in the residential system of education as not to allow any distraction."

"Do you think that there's teachers' more attention to the students?"

"Yes, the teachers in the system are ever present with them,"

"Do the students feel full freedom in the system?"

"Students in quest of knowledge never feel that they lose any freedom in the residential system of education."

"Are the teachers responsible for the failures of the students?"

"As per the psychologists, teachers are ten to twenty percent responsible for the failures of the students."

"Are the teachers responsible for the victories of students?"

"So, is the case with the victories of students."

"Do you say students are responsible for their progress?"

"True...According to Swami Vivekananda, knowledge is inborn."

"What's the necessity of a teacher when knowledge is inborn?"

"The teacher shows the way to acquire knowledge."

"What according to you what is the teacher meant for?"

"The teacher is meant for guiding students. In the system of residential school of education, the teacher always guides the students in the right path."

Like that, Dakshinya Deekshith answered their many questions, and the programme came to an end successfully.

The inmates of the orphanage were happy to see Dakshinya Deekshith on TV. Shasima felt more delightful than others in the orphanage.

* * *

14

Prashanth came to meet Shasima everyday. They spent their quite comfortable time in the lap of nature. They had many love parleys. Their happiness knew no bounds. They proved to be true lovers in all respects.

Shshima waited for Prashanth's arrival for their meeting in the cool bower of verdant trees...All the objects of nature were happy to see them together.

"I'm very happy to see Prashanth and Shasima together in joy," said the fawn by its dark eys.

"We're happy that our lovely girl, Shasima meets the lovable boy, Prashanth," said the cows by their glances.

"The most winsome damsel is with the most handsome boy...The most welcome feature in the lap of nature," said the trees, offering them their cool bower for their meeting.

"Our princess-like Shasima unites with our prince-like Prashanth," said the butterflies, welcoming their union in nature for their love-rapture.

"O Prashanth, our Sushima is as tender as the flowers with its petal-softness. Let your love be so graceful and delightful," said the humble bees by their humming.

"O black bee, you've flown here to enjoy the beauty of the flower," said the breeze by its gentle blowing.

"We wish Prashanth and Shasima a happy married life," said the sun by shooting warm rays.

All invited Prashanth to join Shasima for lovemaking. They met everyday and lived together happily.

Prashanth who came to Shasima every day stopped coming to her one day. She was very much shocked and worried for that. She thought of his coming. Her eyes were glancing in the direction he came towards her. She appealed to all objects of nature in the ways possible to bring her lover to her,

"O moon, you can remind Prashath of coming to me."

"O morning sun, you can light the way to lead Prashanth to me."

"O sky, you can tell Prashanth that I'm waiting for him."

"O clouds, you saw us together. Today we aren't together. Kindly tell Prashanth not to forget me, his wife, Shasima."

"O garden, you enjoyed our union beneath the bower of trees. Prashanth has forgotten to come to me. Send a message to Prashanth, informing him to come to me."

"O parrots, fly to my husband, Prashanth to tell him to come to me."

"O butterflies, I'm sorry that my life-partner has forgotten me. Inform him to come to me."

. .

"O, doves, love's slaves! Fly to my partner in his region on your fast flight and inform him that I'm eagerly waiting for him. I can't tolerate any sparation between him and me, us as lovers."

"O breeze, blow to Prashanth to whiper into his ears that his beloved isn't ready to take even water till he comes to me."

"Waterfalls, my everyday friend, you can let your roar take my appeal to Prashanth not to lose a second in coming to me on his wings."

Shasima looked at all objects of nature with appeal to them to see that her husband, Prashanth comes to her everyday. Her dark eyes seemed to ask them to tell her the reason for his not coming to her.

The fawn seemed to share her feelings, by fondling her. All the cows felt sorry for Prashanth's not coming to her for lovemaking. They seemed to console her in times of her separation from her lover Prashanth.

All objects of nature stopped to glow, as Shasima was not happy because her husband did not meet her.

Shasima was unhappy because her life was gloomy. She did not know what to do. She suffered a lot. God and all objects of nature knew her suffering.

Her father Dakshinya Deekshith came back after his visit to Kolkata and learnt that Shasima was married to Prashanth. It was her love marriage.

Shasima was with her fawns looking eagerly for Prashanth in the direction he came to her previously. He dreamed of his arrival for their union everyday.

For some days, Shasima and Prashath lived happily after their love marriage. Shasima became pregnant. The inmates of the orphanage celebrated the occasion in a befitting manner at the orphanage. Dakshinya Deekshith seriously thought of his responsibility in sending his daughter Shasima to her husband, Prashanth's residence. He at the same time thought of his responsibility in performing the marriages of his daughters: Suma and Sushma soon.

Shasima and the inmates of the orphanage did not know the reason why Prashanth did not come to Shasima. He in fact forgot to come to see Shasima for two months. Shasima did not know the actual reason for his not coming to her. She did not even know what happened to him. She prayed to the goddess Srivani everyday to send her husband Prashanth to her.

... ...

. Prashanth forgot to come to Shasima due to the plot of Smita and her father, Jagan in his village, Vishvanagar.

Smitha, a daughter of a rich magnate developed secret love for Prashanth as he was good looking. She wanted to achieve his love for her by hook or crook. When she failed to achieve her goal, she told her father, Jagan, that she was going to put an end to her life.

Smitha reiterated that she was in deep love with Prashath, adding,

"I can't live without Prashanth. He isn't responding to my love for him. I want him to marry me, my dear father."

"Have patience. I bring him to you ... to marry you," said Jagan in a whiff of anger.

"Prashanth should sit beside me, walk beside me, breathe beside me...He should look at me appreciating my charms twenty-four hours a day."

"You know what your father, Jagan is... Rest assured...," said Jagan, leaving on the formidable and inevitable mission.

Jagan was a short-tempered man. He felt that it was his prestige to achieve his goal soon under any circumstances. His daughter's mission was his goal. He took it as a challenge to please his heart with pride by fulfilling the love goal of his daughter.

On enquiry, Jagan learnt that Prashanth had secretly married Shasima. It was their love marriage. He was living in the sweet memories of his love for her. He made up his mind to force him to marry his daughter, Smitha. He promised to offer him all his property. Prashanth was not ready for that.

Jagan decided to make Prashanth forget the sweet memories of his past. He expected Prashanth to marry his daughter Smita without fail then. He was on his mission to please his daughter, Smita very much and all possible efforts to reach his goal.

Prashanth lost his memory of the past, especially of his love marriage, as Jagan was successful in administering a drug to him in a party to make him forget his past, his love for Shasima. As a result, he lost his memory of his past. He forgot the event of his love marriage with her in the past. He lost his smart phone with his love-marriage photos somewhere.

Prashanth became sick after having the drug administered by Jagan in a party.

Jagan saw Prashanth, admitted in the hospital to recover from sickness but not to regain the memory of the past. Prashanth was undergoing medical treatment under the supervision of a super specialist.

Jagan rushed to his daughter Smita in happiness and said in eagerness, "Congrats, Prashanth is recovering from sickness. Soon after his recovery, he will marry you."

"You're my father. You are unlike the fathers of others. Mr. Jagan is my father. My father is great," said Smitha smiling.

Prashanth was very famous for belonging to the family of Vishwanath who was famous for farming and crop production in Vishvanagar. He was the only son. He was very handsome to attract others. He had super leadership qualities. All loved him in their hearts.

Smitha saw Prashanth in her school days. She fell in love with him at first sight. She made up her mind to marry him by making her love successful.

“By hook or crook, I must win Prashanth as her husband,” she thought,

"My life will go waste if I don't marry Prashanth. I'm waiting for that precious opportunity."

Jagan was fond of his daughter. He wanted to fulfill her desire somehow or the other. Her sole goal was that she wanted to marry Prashanth.

Smitha dreamed of marrying Prashanth and becoming the rightful owner of his father's properties and estates in Vishvanagar. His father was not an ordinary agriculturist. He had orchards of various types. He had paddy fields in two hundred acres of land. He had a good name as a farmer. He was the farmer of farmers as they sought his advice in farming. She dreamed of becoming a daughter-in-law in his family.

Smita started to dream of marrying Prashanth soon after his recovery. She fell in deep thoughts one afternoon. She had all expectations about Prashanth's positive responses. She fell asleep with all expectations and had sweet dreams.

....

"Hello, my darling, I learnt that you fell in love with me in your school days," said Prashanth in the dream, while his dimples glowed brightly.

"That's what I wish to tell you, my dear," said Smitha.

"I love you for your love is pure. You're in the devotion of love," said Prashanth.

"I've adored you since my school days," said Smitha.

"I'm glad that you've adored me since your school days," said Prashanth.

"You can ask my heart. It'll tell you about my love," said Smitha.

"That's why I have come to marry you," said Prashanth.

"My dream has come true. My lover has come to me to marry me. My father's plan is successful...," said Simitha.

Jagan also appeared with all smiles, as he was successful in his plan. He brought garlands for Prashanth and Smitha to exchange. They exchanged their garlands. Their marriage was over as per the will and wish of Smitha and her father Jagan.

"Congrats, Smitha," said Jagan.

"Thank you... daddy. I profusely thank you, my dear daddy. I've won Prashanth as my husband. I'm the daughter-in-law of Vishwanth, the nation-famous landowner in Vishvanagar, " said Smitha happily.

Prasanth and Smitha sang a duet on the eve of their marriage. They felt on top of the world. She cried highly in her utmost happiness.

Jagan rushed to his daughter, Smitha, when she shouted in happiness in her bed one afternoon when asleep. She had a dream as per her dreams. She woke up from her sleep suddenly and found her sweet dream shattered.

"What happened, my dear Smitha?" said Jagan in surprise.

"It was a dream, not a reality," said Smitha, shedding tears.

"Don't worry. Your dream'll come true...It's my plot with my successful plan," said Jagan confidently.

Smitha was sure of marrying Prashanth after his recovery. She kept waiting for that precious day.

* * *

15

Shasima was leading her life in the great pain of separation from her husband, Prashanth. She was not able to bear that separation. Dakshinya Deekshith made up his mind to send Shasima to her husband, Prashanth,

"The woman married must be with her husband only," said Dakshinya Deekshith.

They enquired about the place of Prashanth. He learnt that it was Vishwanagar. It was famous for farmers and their agriculture.

Shasima learnt that Prashanth from Vishwanagar. She was happy to know the place of her husband. She wanted to reach the place by hook or crook. She did not know her husband's place nor did her husband inform her about his place, Vishwanagar. Their love was love and they were the lovers who needed no introduction for love is like love. Lovers are like lovers. Lovers are pure like lovers and their love is true to be flawless like love. They never felt that love never wants to know the addreesses of lovers' places, the statuses of their families, the sociological backgrounds of their families and so on. Love blooms like a flower. Its location is not important. Its gardener is not important. The love of Shasima and Prashanth bloomed like a flower for its fragrance and excellence. Some circumstances made her know her husband's place and position, his state and situation and his clan and condition and so on when the lover was to forget his wife due to some drug administered to him.

Shasima was happy to know the address of her husband Prashanth. She got ready to see him soon and sought permission from her father Dakshinya Deekshith in that regard. She sought his blessings. She hugged him, shedding tears on leaving him. He kissed her forehead in utmost affection for his daughter with his profuse blessings.

In her speedy getting ready, Shasima forgot to wear her wedding ring. She felt all the time that she was leaving her parents and orphanage, sisters and brothers, birds and animals, fawns and calves, plants and trees, butterflies, and humble bees. She was sorry to leave the lap of nature. Her tears knew her love for the objects of nature and the members of the orphanage.

Shasima along with her mother Annapurna and her father's disciple, Raghuram went to her husband's palace. She went on recalling Prashanth's love for her and their love marriage. They were in deep love for some time. Time went on happily. When she recalled their love to break suddenly, she kept shedding tears continuously.

They were on their way to Prashanth's village, Vishwanagar. They earnestly reached the place.

Shasima learnt what had happened to her husband Prashanth and so she approached the hospital where he was taking medical treatment. She did not know the reason why he was to take medical treatment.

Annapurna, Raghuranm and Shasima wanted to see Prashanth in the hospital. The hospital officials did not allow them to see him. To see him in the hospital, Raghuram managed to get permission with the help of Prashanth's close friend Prakash. They entered the hospital to see Prashanth and felt sorry.

"Prashanth, your Shasima has come to see you," said Shasima to Prashanth.

Prashanth looked at Shasima. There were no feelings in him on seeing her. He was not able to tell her who she was. He did not try even to recall his loving and marrying her in the past. They all felt stunned.

The situation prompted Shasima to remember the ring, their love token and show it to remind him of his love marriage with her. She looked at her ring finger with the ring. The ring was missing. She instantly felt like asking him to search for their love marriage photos in the gallery on his mobile smart phone, but it was not with him. She said to all the people present there,

"I'm Shasima, Prashanth's wife."

"Prashanth loved and married me."

"We're couple...Prashanth is my husband...I'm his wife...We married in the presence of the sun and all objects of Nature. They witnessed our marriage."

Shasima went near Prashanth and spoke to him. She tried her best to make him recall his marriage with her in all the ways possible.

"I'm the girl, Shasima, you saw in the orphanage."

"You fell in love with me at first sight... I'm your darling... Our love culminated in our love marriage. We met in our wedding in the lap of nature near the orphanage..."

"I'm your wife...your life partner."

"I was playing with the fawn in the orphanage when you first came to me."

"The orphanage is near the waterfalls you visited along with your friends."

"You enjoyed the beauty of waterfalls near the orphanage."

"You saw Suma and Sushma, my sisters on the premises of an orphanage."

"It was at my father, Dakshinya Deeksith's orphanage in the lap of nature."

"The humble bees and the butterflies are my friends, you've seen them, hovering round me."

"You saw the calves and the cows with me to fondle."

“I was playing with my fawn in joys at the time of your visit to me.”

"I smiled like this, bowing my head in shyness when you expressed your love for me first."

"You don't remember our love marriage. We were very happy together in the cool bowers of trees for some days. Our happiness knew no bounds. I remind you of it. This is a reminder for you to recall it...Your picnic to the waterfalls. You love me at first sight...My love to you...Our wedding ring with your name and my name...wedding ring with our miniatures in it...Our photographs in your smart phone..."

Prashanth was not able to recall all although she told him about the past incidents. When his face looked pathetic, Shasima sat close to him. She looked into his eyes in love and affection. She kissed him in grace.

It was not possible for Prashanth to recall Shasima. His face was sans responses and expressions.

The people, the near and dear of Prashanth noticed her to be pregnant. She was carrying a fetus in her womb. His friend Prakash who was aware of their love affair was helpless in telling his friend the needful. He did not know what to do and how to convince the people there. All kept silent with their unheard feelings and unseen views. They were mute observers of the scene.

Meanwhile, the hospital officials asked them to quit the room soon. They had no way, except to go out of the room. They quit the room feeling fully discouraged and came out in disappointment.

Still Shasima had all hopes that Prashanth would surely recognize her as his wife. She was eagerly waiting for abundant chances to speak to him. It would be the most precious occasion in her life.

"No woman says that some other man is her husband and so Shasima is true to her expressions and feelings that Prashanth is her husband," all the people thought.

The well-wisher of the family, Anand had a soft corner for Shasima. He advised her to wait at the village Vashwanager until Prashanth recovered. He also requested her to stay along with her mother Annapurna and her father's disciple, Raghuram in Prashanth's family guesthouse.

Annapurna and Raghuram along with Shasima stood beneath a tree, thinking of the ways to reach Prashanth in good condition. They felt that it was not possible for them to see him as they learnt that it would take some more time for him to recover fully.

Annapurna, Raghuram and Shasima felt that it was not proper for them to stay in the family guesthouse of her husband, Prashanth when he was not able to recognize Shasima as his wife and life-partner. In such circumstances, they took a decision to leave the place, Vshwanagar.

"Let's go to the orphanage of Harsha Deekshith," said Raghuram to Annapurna and Shasima.

Raghuram and Annapurna along with Shasima were on their way accordingly to the orphanage of Harsha Deekshith who was his father

Dakshinya Deekshith's close relative. They were intimate to each other as well.

They undertook their journey to see Harsha Deekshith at Harsha orphanage, Raghuram narrated all in their journey what had happened to Prashanth as per his expectations and observations,

"A great injustice was done to Shasima. Smita loved Prashanth but Prasnanth didn't love her. That was one-sided love."

"Smitha wanted to achieve her goal of love at any cost. She told about her love to her father, Jagan, a man of villainous nature. He plotted to administer a dose to Prashanth to make him forget his sincere love for Shasima."

"Prashanth became a scapegoat...He became a victim to the evil goal of Jagan."

"In administering the dose, Jagan mainly aimed at making Prashanth forget his love for Shasima. He wished him to recover from sickness without gaining the memory of his love for Shasima, and marry his daughter, Smitha."

"In fact, Prashanth is in love and in the memory of his wife Shasima. There is no doubt about it..."

"That was the cold-blooded plot of Jagan."

Narrating all the plot of Jagan in making his daughter, Smitha's dream of marrying Prashanth come true, they journeyed to the Harsha orphanage.

Annapurna, Shasima and Raghuram arrived at the orphanage. Harsha Deekshith's wife, Sahridi received them with all love and affection, calling Shasima by name.

"You're calling me by my name with all your love like my mother. How do you know my name, my dear mother-like mother?" said Shasima, feeling relieved at her hearty call.

"Dakshinya Deekshith is more a friend than a relative. My friend's daughter is my daughter," said Harsha Deekshith affectionately while Sahridi was nodding her head in full acceptance.

"Shasima will stay with you for some time. We all feel that to stay with you at the orphanage nestled amid beautiful nature, is very suitable. She can spend her time with the fawns and calves around it as she did at the orphanage of my friend, Dakshinya Deeksith. She can live here happily forever," said Harsha Deekshith.

"Most welcome, she's more than our daughter. She'll be here with us," said Sahridi, the wife of Harsha Deekshith happily.

With heaved hearts, Raghuram and Annapurna went back to their orphanage of Dakshinya Deekshith, leaving Shasima in the orphanage. They informed Dakshinya Deekshith on the phone that they were coming back to their orphanage, leaving Shasma at the orphanage of Harsha Deekshith.

....

Harsha Deekshith introduced his orphanage to Shasima in affectionate terms:

"This is the orphanage of Harsha Deekshith. It's like your father, Dakshinya Deekshith's orphanage at the outskirts of the forest."

"I'm your father-like, more than your father...It's your orphanage. You're free to live here as my daughter. "

"I too have cows in the orphanage. You've calves for you to play with them. You've fawns, the young deer also for you to fondle them."

Harsha Deekshith in the conversation reminded Shasima of her fawn at her father Dakshinya Deekshith's orphanage. She also remembered her fawn, the cows, plants, and trees there and smiled.

"Here the wild animals come to graze and then go back to the middle of the forest... They don't harm the inmates of the orphanage."

"Very rarely cobs are seen. The inmates of the orphanage love them."

Shasima found all comfortable to live there and so she felt relieved of the pain in her heart when Harsha Deekshith showed his orphanage with all fatherly love and affection to her.

Shasima lived in the orphanage and slowly recovered from her shock on Prashanth's forgetting her as Harsha Deeksith and Sahridi treated her as their daughter.

16

Shasima was in the lap of nature enjoying its sights, sounds and scents. She was able to forget her feelings on her separation from her husband Prashanth.

Shasima was happy before her delivery as she found all the lovely things in the orphanage of Harsha. The inmates were very affectionate to her. In that happiness, she was waiting for the golden day for her to deliver a baby on an auspicious occasion.

The inmates of the orphanage looked after Shasima with love and affection, care and concern. She never felt that her mother was not with her.

Harsha Deekshith gave instructions to all the inmates of his orphanage, especially his wife to take special care of Sahasima in the days for her delivery.

On the most auspicious day, Shasima delivered a male baby in the orphanage. She was very happy when she gave birth to a male baby. The baby was very handsome. It looked charming and dazzling. It shone like the morning sun in charms. It at the same time looked daring and dashing.

The male baby glanced all, turning its looks in all directions and drawing their attention with its tender smiles and graceful movements.

The inmates of the orphanage were very happy to look at the baby with winsome qualities and handsome features. They felt it was the rarest fest in the history of the orphanage.

Harsha Deekshith wanted to celebrate the occasion in a befitting manner. He thought of the most suitable name for the baby, looking at the boy and his cuteness and sharpness.

Harsha Deekshith named the baby, Bharath wishing him that he should become a great leader to rule India, establishing peace and harmony in the nation. He was in the lap of every woman present there on the occasion and pleased all by his tender smile.

Harsha Deekshith had a chance to observe the heroic qualities in the budding stage of the newborn male baby and so he was overwhelmed with joy.

When a flower starts blooming, it spreads its perfumes and shines in all directions for the pleasure of everyone. So is the case with Bharath.

Harsha Deekshith wished the baby that he, as a leader one day, should treat everyone with love and affection like the sun to shoot its delightful rays and the flower to spread its blissful treasures.

All the women were present on the occasion. After the naming ceremony, they blessed the baby in their own ways:

"The baby is smart. He'll be a great leader."

"Wow, you aren't ordinary like others. You're different. You'll win the praises of all the people."

"You'll excel your father in all respects. You'll be unrivalled."

"You shine like the sun. There is an incomparable charm in your face. All love you very much, I 'm sure..."

'Your mother is fortunate to have you as her son. She'll rise to solve all her problems when you grow up."

"Your mother has faced a lot of torture. You've come now to see that your mother is very happy."

"Your mother knows only success."

His mother, Shasima was in her full smiles. He looked at her son with all joy. She took her son into her arms. She expressed her joys:

"The clouds that hid my glows of joy in my separation from my husband will be shattered soon. Happy days are ahead. I've been waiting for your dawn. You've dawned now. Happy days are fast approaching us, my dear Bharath. I wish that you would shine as an ideal leader and turn a model to the people. You should rise to national and international fame."

Shasima laid Bharath in the swing. They just touched the swing. It started to swing on its own as a wonder.

All the women circled round the swing with the baby in swing and sang a song showering their blessings and benedictions to Bharath:

Refrain: We all bless you to be a victor

We all bless you to be a victor.

Stanza: We all bless you to grow human

To render your services yeoman

To fill all hearts with full mirth

In all the lives born on the earth. //We all bless//

Stanza: To our nation, you are a treasure

To offer joys in the full measure

You are dawned for their dream

To sail in the blissful life-stream. //We all bless//

Stanza: For the human clan, be a leader

For a good cause, be a crusader

You are the sun to shatter vices

You are the gun to shoot crises. //We all bless//

Refrain: We all bless you to be a victor

We all bless you to be a victor.

All participated in the function happily. Shasima's happiness knew no bounds. All the members of Dakshinya Deekshith sent their heartiest greetings to Bharath on the happy occasion.

....

Bharath grew to be a tiny boy of attractions and impressions. He resembled his father, Prashanth. He grew to be a four-year old boy. He grew to be a lover of nature. He looked at trees full of flowers and animals and smiled tenderly.

When days passed, Bharath looked at cobs, he felt fascinated to play with them. He started to play with cobs, fawns, and calves. He fought with cobs to prove himself to be stronger and braver than the young animals.

Shasima was extremely happy as she found her son very bright and brilliant, daring and dashing like his father, Prashanth. She was confident of her son, Bharath's success in all fields. She felt full relief with the birth of her son Bharath in her life in the days of her separation from her husband, Prashanth.

Bharath's happiness knew no bounds in his play with cobs. He felt at ease playing with the cobs like a warrior. Others felt somewhat scared at the sight of cobs, but he did not feel scared at the sight of cobs. He moreover played with them, going near them.

....

Days passed for Prashanth to recover from sickness and recall the past. He thought of Shasima all the time. The driver of his car brought his smart mobile phone, saying that he got it under the car seat.

Prashanth smiled at the sight of his mobile. He took it and operated on it. He saw all his photos with Shasima in the gallery. He was very happy. He smiled very often when he looked at his mobile. He saw the pictures of his love-marriage with Shasima. He recalled all the happiness connected to the auspicious occasion of his love marriage in the lap of nature.

Prashanth felt sorry for not knowing what had happened to him recently. He heard through others that he had not recognized her as his wife. He felt sorry for that.

Prashanth thought of Shasima later. He was in the mood to ask others to find out where she was and how she was.

Gradually Prashanth felt better day by day. He was recalling his past and smiling to himself. It was the most pleasurable improvement in life.

Shasima's sisters, Suma and Sushma came to visit Prashanth at his residence in Vishwanagar. He was able to recognize them and call them by names. He recalled those times when he saw them for the first time in the surroundings of the orphanage of Dakshinya Deekshith.

"When I met Shasima for the first time, you were standing at a little distance from us," said Prashanth.

"Yes"

"It was a pleasurable experience for me to meet Shasima daily. It was the venue for our love-marriage. I can never forget it," said Prashanth.

"Not possible..."

Prashanth inquired with them about Shasima's well-being.

Suma and Sushma were glad that Prashanth was able to recall the sweet memories of his love marriage. They expressed their feelings,

"We're sure of your reuniting with Shasima...The thing is that you've to wait for the precious day to start living happily as a model couple," said Suma and Sushma, feeling sorry for Shasima for not possessing the wedding ring when she was in the mood of showing the ring to Prashanth.

"Thank you for sharing my feelings," said Prashanth.

"The wedding ring isn't with Shasima," said the two sisters.

"Where is that wedding ring?" said Prashanth.

When Prashanth eagerly wanted the ring, they gave it to him. He looked at it and tears welled in his eyes in joy. He recalled the sweet memories connected to his love marriage with Shasima on looking at it. He wept bitterly for missing Shasima, and they consoled him, saying,

"The happy moment of your reunion is very near...fast approaching."

Suma and Sushma left in happiness for their sister Shasima staying at Harsha Deekshith's orphanage.

17

Jagan was in touch with the super-specialist doctor to make his plot successful as he aimed only at making his daughter, Smita's dream fulfilled. He enquired about the health of Prashanth.

Through the super specialty doctor, Jagan learnt that Prashanth had recovered from his sickness. He felt that his plan would be successful. He thought that he would perform his daughter, Smita's marriage with Prashanth soon. In the plan, his happiness knew no bounds.

Jagan approached Prashanth and started to speak to him very cordially. He did not let his plan known to him even a bit, but he expressed his feelings openly,

"Dear Prashanth, I'm happy on the eve of your recovery. I heartily congratulate you on your recovery as per my wish. This is the happiest occasion for me. Let's celebrate it in the most befitting manner," said Jagan, observing Prashanth very keenly.

Jagan expected that Prashanth had forgotten his love marriage with Shasima in the past life on one hand and became healthy to marry his daughter, Smitha on the other as per his long-cherished wish.

Jagan was happy and thought of how to proceed to be successful,

"Jagan recovered from sickness, forgetting his love marriage with Shasima in the past. This is the most welcome gesture..."

"If at all he recalls his marriage with Shasima, what shall I do?"

"No, he won't recollect his marriage with Shasima...I'm sure...This won't happen as per my wish, expressed to the super specialty doctor..."

"The doctor is good at special diagnasis and specific treament that helps Prashanth to recover from sickness but not to regain the memories of his love for Shasima."

Jagan was very confident in that and so he wanted to refer to the marriage of his daughter, Smita with Prashanth, saying to him in a humble manner,

"As per the wishes of my daughter, I booked a function-hall for her marriage."

"Prashanth, today it is the most auspicious day for the occasion of your marriage with my daughter, Smitha."

"Smitha is ready as the bride. When you get ready as the bridegroom, we go to the function hall soon for your marriage."

"My dear, get ready for your marriage on the auspicious occasion."

Prashanth remained calm and quiet, and slowly thought of his life partner Shasima all the time. He felt sorry for the gap created temporarily in his married life. He recollected his wife, Shasima, and the memories connected to his marriage with her. They were haunting him. He felt anxious to see her soon. He said to himself openly,

"Shasi... Shasima..."

Jagan noticed Prashanth recall his love marriage with Shasima in the past and felt shocked. He heard him recalling,

"Shasima, I'll definitely reunite with you. The occasion of our reunion will be happier than that we had on our union at the spot where we had met each other for lovemaking."

"It'll be our reunion more memorable than our love marriage. Now our reunion is like the communion of the two rivers to flow to the ocean of bliss in our life."

Jagan recalled what he said to the doctor, "My wish is that you should cure Prashanth's sickness only."

The doctor gave assurance to cure Prashanth's sickness. Jagan was very sure of his treatment to help him cure only his sickness but not enabling him to recall the memories of his love for Shasima and love marriage with her as per his plan. The treatment helped him regain the memoirs of the complete past against his wish.

Jagan always found Prashanth recalling his love and his happy love-marriage with Shasima. He felt extremely sorry for his plan failed totally.

When Prashanth recovered from sickness, Jagan was happy, but he was not happy for Prashanth to gain his memories and help him to recall his past, his love marriage with Shasima. He was angry about that. He had no way except to distract Prashanth's attention to Shasima.

Jagan had his own plan to abduct Prashanth to enable him to marry Smita.

Smita appeared as the bride well before Prashanth was ready for her marriage to him. All the members of Jagan's gang were ready to abduct him by force. They were to take him with Smita for their marriage.

All were ready to execute their plan successfully. Before that, Smita made her own efforts to impress Prashanth and so she started her conversation with him,

"I love you very much. You don't know when I started to love you."

"Since I was a school-going girl, I've been in love with you, my dear, my life partner…my better-half."

"I adore you as my lover... my would-be husband."

"No... You're not my would-be husband...You're my husband...You became my husband long ago. We are wife and husband, a coule for the life of our own ... Let's proceed with our marriage for the sake of formality. I know that marriage is for the sake of formality. Let's be ready for that."

"Let's be united today in our marriage to lead the life of our own in a happy manner as a couple to be model to others."

"The astrologers have already said that I'm fated to marry you."

Prashanth kept quiet and remained calm to all the developments and comments. He opened his mouth to speak to her for the first time.

"I married Shasima. All the people here have known her as my wife since my love marriage with her."

"Shasima is my wife... my better half…my life partner. We are two in one. I'm going to reunite with her."

.

Smita heard Prashanth's views silently. She responded, speaking to him openly:

"I'm your beloved. I adore you as my lover. Where should I go, leaving you to somebody else?"

"Love isn't one-sided and lopsided, double-edged and trouble-edged," said Prashanth unhesitatingly.

Prashanth expressed his love for Shasima, saying to himself very openly,

"Shasi...Shasima... Do you think that I can live without you? It happens never... You're my life...You're my breath...You're my heart... I can never forget you again... Never after..."

"Shasima died long ago...," said Jagan suddenly.

"Shasi, you can't die... You never die...You live in my heart," said Prashanth.

To execute the secret plan of his action, Jagan called his gang, whistling. The gang was ready to abduct Prashanth to the function-hall for his marriage with Smitha. He and his gang thought that it was the only way to abduct him as per their plot.

The members of the gang started to abduct Prashanth. He voiced to protest their plan. In support of Jagan, they started to fight with him, while trying to abduct him. He fought very bravely against the gang.

Meanwhile many people joined to support Prashanth and fought for him, teaching a bitter lesson to the gang.

All the supporters of Jagan ran panic-stricken. None of his gang was there seen on the scene. Smita disappeared from the scene, saying,

"You can never understand me, my sincere love for you, Prashanth. In ignorance, you're shattering my dreams in making our love successful and fruitful."

Jagan looked in all directions. The supporters of Prashanth were after him. Then he took to his heels without turning his head.

Jagan felt stunned as he failed in all his efforts despite his most possible efforts to make his plan successful.

The remaining people forced Prashanth to marry Smita. They threatened him by showing knives. His supporters chased them immediately. Then they ran away as they felt all hopes lost.

Prashanth was brave enough to fight against Jagan. To help him, his supporters came to him timely.

Jagan's supporters were in the wrong way to force Prashanth to marry Smita although they knew that he had already married Shasima.

Moreover, Jagan brought the demon-like-villains with weapons to make his plot successful but tasted defeat and failed to succeed in his plan.

* * *

18

Prashanth had completely recovered from sickness. He started to think of his reuniting with Shasima. Unexpectedly he heard that farmers in an area were facing problems. As a supporter of farmers like his father Vishwanath he went to the rescue of all farmers in the region. He learnt that the lands of farmers in the region were grabbed by some cheats. The lands were ready to become the sites for the construction of houses in a mega city as per their plans.

There was Adarsh, Prashanth's follower, to protest all that in the region. He tried his best to solve the problems of the farmers, but he was not successful. He wanted the help of Prashanth in that.

Prashanth was in the inspiration of the model farmer, Vishesh at Karshaka Kshetra for sharing his lands to the small farmers of their lands. He too had high regard for daring Malli at Indrapur for sharing her land to her workers in her agriculture field.

Adarsh was very sorry for the situation when all agricultural lands were to be turned into sites for the tenement of houses to come up everywhere with all rich amenities and luxury facilities in the farmers' lands of green fields. He felt very much sorry for he envisaged a death blow to the industry of agriculture in the region and expressed his feelings very often,

"Ours is an agricultural country. It's the key industry of India. It's famous for that across the world."

"We see evergreen crops for food and verdant trees with the cool shades for the life of the people in India. The green color of our national flag represents our green wealth and rich crops. We are very proud of our nation for that."

"All the lands will turn into houses for cities with industries and factories to cause pollution. All the lands for natural resources become cites for houses with lavish luxuries and cozy comforts for transient pleasures."

"This shouldn't happen. I'm against all this. For the vested interests, all the Real Estate Businessmen are planning to misguide the farmers saying, 'agriculture is waste'."

"Agriculture should be there...farmers should be there...green fields must be there...We're meant for all these as they are there to say that agriculture is the key industry of our life... the way of life... the culture of life...the life of our life."

Through Adarsh, Prashanth learnt all about Abyuday and his illegal grabbing of lands and reacted against him, saying,

"Abhyuday is notorious for his illegal occupation of lands. He says that it's his business... fair business. Business in fact is the way of life. It's a profession to get reasonable profits for livelihood but not losses to the customers especially the farmers to produce food grains in the region. Nowadays business is the means for one to rise to be a billionaire causing loss... deception... exploitation of the farmers. He should realize the ethics of business."

Abhyuday as a businessman was ready to grab all the lands from the farmers in the region with a plan to build a big city in the agricultural lands for the dense population to live in it. He cleverly exploited the innocent farmers in the region, telling them cock and bull stories.

That was the situation prevailing everywhere. Prashanth thought of the safe existence of the farmers in the future generations. He decided to join Adarsh, the well-wisher of agriculture and the staunch supporter of farmers and help the farmers.

Meanwhile, Adarsha made a call to him to Prashanth in full tension,

"Hello Prashanth"

"Hello Adarsh,"

"I draw your active attention to the pathetic situation of the farmers in the region. Here all the lands are ready to become sites for building housess with houses for the farmers to live in, Abhyuday misguided them by saying that they'll enjoy luxuries in the houses to come up in their lands," said Adarsh.

"That should not happen at any cost. I'm rushing to you to see and teach Abhyuday a bitter lesson. I know such people deceive farmers. The people like Suryasen in the past and those like Indrasen in the present proved to be cheats of the famers community. Such people should learn a lesson that farmers are for farming as the farmer is the backbone of our nation...It is certain...

I make Abhyuday realize the importance of lands for agriculture, the key industry of our nation," said Prashanth.

"Here all the farmers in the region are the losers. They'll become landless farmers. Abhyuday is the deceiver and exploiter behing this evil plan. You've got to teach him a bitter lesson," said Adarsh.

"They get back their lands from the land grabber and continue to own and farm the lands to turn the region fertile and verdant as usual," said Prashanth.

Prashanth was for the welfare of farmers. He did not tolerate any harm done to the farmers. He was sorry for he heard that the farmers were becoming victims to Real Estate Businessman and grabbers, Abhyuday in the region. He went to rescue the farmers from being deceived. He took much risk to make them free from his shackles of deception and manacles of exploitation.

'Abhyuday' was the choicest name for his father, Dharmaraj, to name him Abhyuday wishing him to have rational thoughts and democratic temperaments and be worthy of the name. He wanted his son to have a good name in society, but he was a rogue to learn a bitter lesson.

In life, Abhyuday instead turned out to be a man to grab the lands of innocent farmers against the dreams of his father, Dharmaraj.

Abhyuday was an ordinary merchant once. He was not happy with his earnings. He lived in a thatched house then. He felt that it was not comfortable for him. He dreamed of constructing a big mansion. He at the same time craved a white-collar job, but it was not possible for him to study in an institute. He found all the subjects hard for him to study. He tried to make all his subjects easy somehow or the other but failed to do it miserably. When he found studying was very difficult, he gave up the idea of studying for a white-collar job.

Abhyuday thought of the easiest way to rise to the level of a multibillionaire. He chose the job of a Real Estate Businessman and land grabbing for that purpose. He slowly tried to grab farmers' lands by means of his evil plans.

Prashanth went to the region and studied the situation of the farmers in quest of solving their problem. The farmers in the area were very innocent. They were work-cultured and profession-committed, but they were unaware of the ways of Abhyuday to cheat them.

The farmers of the region blindly believed in Abhyuday who dealt with them in the sale of fertilizers, pesticides, seeds and so on. He deceived them in the sale of all these. They sought loans at interest to purchase all these.

Abhyuday had appeared to them to be a man of helping nature and the well-wisher of farmers. He told all the farmers that he was there, coming to their rescue. He was their godfather. They easily believed him to have his helping hand.

The main profession of Abhyuady was money lending to grab their lands when they failed to repay loans. The farmers borrowed loans for different purposes also from Abhyuday. Whenever they borrowed loans, he asked them to mortgage their lands, saying with a cordial smile,

"I've full confidence that you'll clear the loans. I want you to sign the documents only to remember what you have taken from me. You know that I'm your near and dear... your well-wisher and your benefactor."

The farmers signed the documents without any hesitation, without seeing the conditions mentioned in agreements. They expressed their views in terms of close relation and cordial concern,

"When you're our well wisher, you're our protector and benefoctor. You're a man of kind heart and benevolent soul for the backbone of the nation. When our lives are in your hands, our lands are in your hands. Our lands are your lands as you look after us, our agriculture."

Abhyuday was very clever to have created an irrevocable faith in the minds of the farmers. He was able to remember the names of all their lands. He mentioned in the documents signed by them.

There were times to ask Abhyuday, "How can you remember the names of our lands? You feel at ease in telling the names of lands."

As a surprise to the farmers, Ahyuday once told the farmers all the names of lands in Telugu: 'Moddula Paya', 'Tangella guddam', 'Pedda Ragedu', 'Balaraju Peradu', 'Kona Polam', 'Pochamma Gadda', 'Pilli-gundlu', 'Regadi Cheluka', and so on like the priest to recite the names of God in his hymns. He knew all well who the owners of the lands were and that was his sharp memory though he failed in all subjects at school due to poor memory.

The farmers felt surprised at his sharp memory in telling the names of lands. Whenever they asked him about his study, Abhyuday gave them a sincere reply to make them think of him highly,

"Abhyuday, M.A., B. F., I.A.S."

Abhyuday was earlier such a student that the people of the region innocently thought that he had studied a lot. They felt that he was the man that God had sent him to look after them, the innocent farmers.

The farmers understood that Abhyuday had studied a lot to understand their problems, troubles, and difficulties.

The people with a little education thought of his M.A. It was his post-graduate degree. They thought of his B.F. It was his new and great qualification. Then they thought of his IAS. It was the qualification for him to become a district collector. But he did not opt for the job, only to look after the people as their well-wisher.

The farmers were under a good impression about his study for a long time. In the years to come, they overheard what his qualification was indeed.

Abhyuday came to the Matriculation class as there were no examinations in the previous classes. He in fact appeared for the examination for the first time to pass matriculation but failed in all subjects as he treated all the subjects equally and gave equal preference and importance to all subjects and so he failed in all subjects.

The farmers never thought that he was **M**atriculation (tenth class) **A**ppeared, **B**ut **F**ailed, **I**n **A**ll **S**ubjects to be abbreviated in bold with the initials of every word as M. A., B. F., I. A. S.

The farmers all learnt that Abhyuday was a clever and cunning fellow. But he was a poor student to have failed in all the subjects at his school final examination for his poor memory.

All the farmers of the region in fact did not know that he would rise to be a miultibillionaire by grabbing their lands and doing a variety of deceptive businesses.

Abhyuday as a merchant supplied seeds to the farmers. Sometimes the seeds supplied by him were hybrid, sometimes spurious and very rarely they were original. They did not know that he was that kind of cheat. The seeds did not germinate. Though germinated, they did not grow for yielding. Some farmers committed suicides throwing their families into streets.

On credit base, Abhyuday readily supplied seeds, fertilizers, pestisides and others on credit basis. He lent loans to them at a high interest for different purposes like the children's marriages, education, tours, pilgrimages, excursions and so on. They happily borrowed loans from him. saying with all smiles,

"Please sign the document to mortgage your land."

"We sign wherever you want us to sign," said the farmers.

“I’m for you. You’re for me,” said Abhyuday.

“Yes…you’re for us…We know,” said the farmers.

The farmers felt very comfortable in taking the loans as hand loans. They never realized that they were to clear the loan amounts with high interest. He pleased them by speaking in a sweet tone, saying,

"This is your money...It’s with me idle. You can take it from me and utilize it. You'll return it. I know that you'll clear it soon. Take this pen and sign the document or affix thumb impression in case of illiterates."

The innocent farmers signed the documents or affixed thumb impressions as they thought that Abhyuday was the wel-wisher of the farmers. He told them all about his master plan in the welfare of the farmers in the region,

"I've a plan to turn you all into millionaires. I use your lands for the sites to build big houses for you, your wife, and your children. You needn't clear your pending loans. You know that you mortgaged your land. You needn't pay any money to me. Instead, you get big amounts for your lands. On the amounts, you get interest every month. I allot the houses to you. Your houses will be with lavishing comforts... You ebjoy living in them,"

"Very glad to hear the good news from you," said the farmers.

"As per this plan, you all get millions. In the region, land of a thousand acres on is available. I'm going to build an A-class maga city. It'll be heaven on earth. In the city, you find not only your buildings but institutions for the education of your children and hospitals for your treatment when sick, the stadiums for sports and games, a university with engineering colleges, medical colleges and all in its purview for your children's education, a pond for the boat-ride, oldage homes with all facilities......You get all facilities...what not... everything... every facility... These will be in your lands in your thousands of acres. You'll be kings and your wives will be queens, living in your royal palaces to come up. Why do you do agriculture, shedding sweat in the scorching sun? You can live in air-conditioned buildings, singing and dancing. You will be like the heroes and your wives will be the heroins like those of the film to sing duets. You needn't enter the agriculture fields and get your hands dust-laden and dirt-stuck."

In respose to his plan in their welfare, they expressed their views,

"You're our godfather...God sent you to us for our welfare."

"You're in fact God decended from heaven to earth for us."

All the farmers felt that Abhyuday was the harbinger of good fortunes for them. When they were to nod their heads, a stranger suddenly appeared on the scene after hearing Abhyuday's master plan and their positive response,

"We build houses with all comforts in the city to come up with all the facilities like university education, hospitals for medical treatment, stadiums for playing and so on in your thousand acres in which you get just food after doing a lot of labor and sweat for that," said the stranger.

"We're happy to live in the city to come up soon," said the farmers.

"Where do you get food?" said the stranger.

"From shops, hotels and other places in the farm of parcels…I'm there for all those," said Abhyuday.

"If all the agricultural lands are turned into cities, where is the land for agriculture? Where is your crop production? Where is your food-production?" said the stranger.

"We can get food grains from other countries."

"If all lands in other countries turn into the sites of houses, there won't be lands for farming. All roadside lands are with boards of Real Estate Business with colorful banners and attractive ads now. Once there were fields to offer us grains for our food. When no farmers are there, there is no food for all to eat," said the stranger.

"Who're you, to talk to us?" said Abhyuday.

"I'm the well-wisher of farmers...Farmers are my friends. The farmer is the backbone of my nation," said the stranger.

"I'm the real well-wisher of all farmers of this region," said Abhyuday.

"You're not the well-wisher of farmers. You're in fact the ill-wisher of farmers. You gave them loans. Now you grab their lands, turning the farmers into beggars. You make them wander in the streets and see them begging," said the stranger.

"I turn them millionaires, taking the lands," said Abhyuday.

"First, think of food, then clothes and shelter; then think of money and luxuries. The primary thing is agriculture to offer us food. We treat agriculture as the key industry of India. The farming is the nucleus profession of our country. You want to break the backbone to make the nation spineless. You turn India lifeless. The plants or the crops that farmers grow are the sign of life. The farmer grows plants and crops for life. It's for our life. It's the way of life. What we eat is the product of agriculture...the fruit of their toil. Saving agriculture is our culture and our responsibility. Don't turn all agricultural lands into sites for building cities," said the stranger with anger in his face.

"When I'm making them multimillionaires, you're coming in my way," said Abhyuday intolerably.

"You're making them idlers," said the stranger.

"When I want them to live in luxuries and enjoy comforts, you are misguiding them to work in lands with sweat faces in the scorching sun," said Abhyuday.

"Don't talk of luxuries and comforts. Talk of food sources for our facilities. The farmers are for those facilities ...for all," said the stranger.

"Who're you?" said Abhyuday.

"I'm the well wisher of farmers to work under the inspiration of Vishesh, the drear of the farmers. Malli is also the farmers to inspire me. I am for the farmers... I'm to champion the cause of the farmers. I am Prashanth, the heat of the farmers. Adhyuday! shut up and leave the spot. You shouldn't live here to misguide the innocent farmers.

“The farmers, clear your loans. Abhyuday, give them time for one year to clear your loans. Let the farmers do agriculture and live on their own like kings. I'll see that your loans are cleared by repaying your principal amounts with reasonable interest."

Abhyuday felt crest-fallen and left the place within no time. All thought that Abhyuday would appear after one year.

19

Prashanth's life passed through ordeals and hazards as his life-travel was full of travails. First, he became a victim to Jagan's plot.

Jagan administered a drug to Prashanth to make him forget his love marriage with Shasima at the orphanage.

Prashanth was under treatment for his recovery. Taking some time, Prashanth luckily recovered from sickness.

Then Prashanth fought with courage to save himself from the evil plot of Jagan. When he was well and free from all adverse circumstances, he undertook his search for his wife, Shasima.

While Prashanth was in the mood of searching for Shasima, he got a clarion call from his friend Adarsh to come to the rescue of the farmers in their region as he championed the cause of farmers.

Abhyuday was ready to grab the lands of farmers by means of his evil plans. Prashanth dealt with him very tactfully on behalf of the farmers.

If Prashanth had not come to the rescue of the farmers in the region, Abhyuday would have grabbed their lands and made them landless and the nation spineless. The farmers would have become landless to be the beggars in the street.

Prashanth learnt that Abhyuday was the gang-leader of the land grabbers. He was supporting the land grabbers everywhere. He had close connections with other land-grabbers. He was the president of the Real Estate businessmen.

Prashanth wanted to teach a bitter lesson to Abhyuday of Real Estate Business, as he was to plan sites for building houses in the agriculture lands of farmers.

That was the situation for the agricultural lands to turn into plots or sites for the construction of houses everywhere as per the plan of Abhyuday and the like. Adarsh tried his best to stop all this.

When everything went beyond limits, Adarsh of the region called Prashanth, the well-wisher of farmers.

....

Prashanth as a well-wisher of farmers smelt Abhyuday's evil plan and appeared on the scene to ague with the evil doer, saying,

"Mr. Abhyuday, you have started your real estate and grabbing business. You want to make use of the lands of the farmers in the region for you to come up. I learnt that they had taken loans, mortgaging their lands. In quest of your goal, you gave them assurance that you would construct houses for them to live in."

"Yes, they live in the houses enjoying their comforts...What's the use of lands and agriculture? They'll have sufficient amount for their happy lives." Abhyuday.

"As I told you earlier, I'm their well-wisher...I'm for the farmers in all the regions not just for the farmers of this region. You're nothing before me."

"I show what I'm," said Abhyuday ready for whistling. Then his gang appeared there with sticks and knives.

They resorted to the means of violence. The fight went on against Pashanth.

Then all the farmers realized the power of unity and united to fight against the gang. The fight went on for some time.

Then the members of the gang ran panic-stricken when the farmers outnumbered them. They at the same felt that they were faulty.

Even after the fight, Prashanth promised Abhyuday to clear the loans of the farmers.

Prashanth had his own plan to clear the loans of the farmers. Somehow, he took a risk and freed them from the clutches of Abhyuday. He got the loans sanctioned for the farmers from various banks to clear the loans of Abhyuday. They did not lose heart to commit suicides. They were confident of clearing their loans under his able and right guidance of Prashanth.

The farmers made up their minds to clear Abhyuday's loans with reasonable interests to be away from his clutches. Then Prashanth appealed to them,

"Dear farmers, you're farmers, the backbone of the nation. Don't believe Abhyuday and such rogues. They're land grabbers. They're deceivers...They're exploiters. They're bloodsuckers. They're everywhere. Continue to do agriculture to be independent for a respectable life. I'll see that loans sanctioned to you at reasonable interests from the banks. Clear the loans given by Abhyuday with the loans you get from the banks."

"You've taught us a good lesson. We've enlightened now. We're indebted to you, dear Prashanth," said the farmers.

"I've solved the problems of farmers everywhere peacefully without making efforts to keep them behind bars. You must be very grateful to Adarsh who informed me about your problems," said Prashanth with all smiles in his face.

One year passed, the farmers cleared off all the personal loans with reasonable interests as per the word given to Abhyuday by Prashanth. They were free from the clutches of Abhyuday. Luckily, the governments waived off the interest they were to pay for the loans taken from the banks. In some cases, they waived off the loans of the deceased farmers when recommended by Prashanth.

Prashanth made the farmers think reasonably and live independently. Credit went to him for his leadership in solving their problems.

The farmers got back the loan documents from Abhyuday. They lived happily farming their lands thereafter.

....

After rescuing the farmers as his foremost responsibility as their well-wisher, Prashanth was on his way in search of his wife. Shasima. In his conversation with her sisters Suma and Sushma in the past, he learnt that she was at Harsha Deekshith's orphanage. He was approaching the Harsha orphanage along with his friend Prakash. Searching for her here and there, he saw the orphanage from a distance.

Coming near, Prashanth saw a tiny boy playing with the cobs and other young animals very bravely. There were fawns nearby. It was very pleasant to look at the scene with the lovely boy.

There was a charm in the face of the boy. The young lad was like him in physical features and actions. He approached the boy thinking that he was his dear son.

"I've missed you since your birth, my dear son," said Prashanth.

"You're not my father. I'm the son of a great man," said the boy.

"Wow, your father is great...!" said Prashanth.

A woman in the Harsha orphanage noticed Prashanth coming to the boy.

"The boy is the son of a great leader of the nation doing wonders. He belongs to the family of splendors," said a woman in the orphanage of Harsha Deekshith.

"What's his father's name?" said Prashanth.

"I don't want to pronounce his name from my mouth. I'm ashamed of ... He deserted the life of an innocent woman, Shasima," said another woman.

"What's your name, my dear boy?" said Prashanth.

"My name is heard from Kashmir to Kanyakumari," said the boy.

"Wonderful...wonderful...," said Prashanth.

Prashanth felt very much impressed by the boy's wit and way of giving replies in cuteness and quickness. He took the boy into his arms and kissed him happily several times. His happiness knew no bounds.

"What's your mother's name?" said Prashanth while Shasima was approaching anxiously to join her son in play.

At the sight of Shasima, Prashanth was overwhelmed with joy. Tears welled in his eyes in joy. He approached his wife, saying,

"I'm immensely happy... I've met my darling, the darling of my heart today. I'm lucky today. This is the luckiest day after my love marriage with you. I learnt about my forgetting of the past and especially you, my darling. I'm sorry...extremely sorry for all that. Forgive me for that, my dear," said Prashanth, coming close to her.

"You're in no way responsible for forgetting me and for my inconsolable suffering. Smita and her father Jagan who were jealously selfish are responsible for our separation...our tears and sighs...our woes and throes," said Shasima with deep feelings.

"I'm happy that you know the reason...It's well and good," said Prashanth.

"I know all, my dear," said Shasima.

Harsha Deekshith joined Prashanth and Shasima with their son, saying happily,

"The past is past. You've reunited. I'm happy for your reunion. Some evil eye separated you not to meet each other for some time. It was due to a secret plot. There is calm after a storm. We've calmness after the unexpected storm...We want calmness and quietness...We've it...Well..."

"I'm very glad to see you, respectable Harsha Deekshith. Kindly bless me, said Prashanth, touching his feet.

"Prashanth, this's your lovely son, Bharath," said Harsha Deekshith, blessing Prashanth heartily.

Prashanth took his son Bharath into arms and kissed him happily and approached Shasima, smiling heartily.

This was Prashanth's most blissful moment of his reunion with his wife Shasima.

Harsha Deekshith spoke to Dakshinya Deekshith on the phone to convey the happy message that Pashanth had come to Shasima for their happy reunion. It was the happiest news in both the orphanages.

"My life was joyless as I was without you Shasima," said Prashanth hugging his wife, Shasima long, knowing no lapse of time.

All enjoyed their reunion when the breeze blew gently. The flowers from the branches above fell with pollen grains on them welcoming their reunion.

Prashanth spoke to Dakshinya Deekshith in intimate terms on the mobile. They were lost in their pleasant conversation.

"I'm to say that I was at fault. Pardon me," said Prashanth.

"There is nothing like that...The past is past...You're in no way responsible for your separation from my daughter, Shasima," said Dakshinya Deekshith.

"I'm happy...I don't have words to express my feelings," said Prashanth.

"True...I'm also in the same situation," said Dakshinya Deekshith.

All were happy in both the orphanges. It was a fest for all of them.

Prashanth took his son Bhrath once again into his arms and kissed the boy when all odds and ordeals he faced as life tides were pleasing him.

They all came to the orphanage blessed the couple on their reunion. Prashanth enjoyed the hospitality of Harsha Deekshith.

Prashanth got ready along with his wife Shasima and son Bharath all together and went to his palace in Vishwanagar. They lived happily so as not to let any misgivings enter their lives thereafter.

Is it their destiny, Smita's jealous nature, Shasima's innocence or Prashanth's yielding to the hostile or deceptive situation that made them suffer, facing adverse life tides for no faults and flaws of theirs?

All happenings became a curse for the couple, Prashanth and Shasima. Bharath was not able to enjoy his father, Prashanth's love.

There was no necessity arisen for the reunited couple to dispel their misgivings. Misgivings did not befall on them, nor did they allow them to occur and recur in their life.

Prashanth reached his place. Prashanth was the son of Vishwanath who was famous for his agriculture in his two hundred acres of land. He treated his workers as members of his family. He was unlike many other landowners who treated his workers as slaves. The people, including his workers, liked him very much and adored him highly.

Vishwanath the father of Prashanth had name and fame and so the people appealed to him to contest the election. He contested the election and won the election with a thumping majority. He was the leader unopposed for a decade in the region, supporting the cause of farmers.

As a leader, Vishwanath solved the problems of the people, especially the farmers, the backbone of the nation. He became a public character. He became very popular for his selfless service. He became very popular as a leader in the nation.

Vishwanath was a farmer first. He became a leader later. As a farmer first and as a leader later he was unrivalled in his region. For him, the father was a leader in a true sense.

When his father, Vishwanth wanted to take a rest, Prashanth took the responsibilities of his father.

Prashanth wished the welfare of all farmers like his father. He went to the rescue of farmers in the region where the farmers became victims to the evil goal of Abhyudaya. He saved them from land grabbing.

Abhyuday wanted to grab the lands of the innocent farmers, telling them that they would become multimillionaires as the owners of houses with lavishing luxuries. He tempted them to offer their lands to him for Real Estate Business.

If Prashanth had not gone to the rescue of the farmers, they would have become landless farmers. He would not have saved them from his deception. Luckily, he went to their rescue.

Like that, Prashanth was the leader to solve the problems of farmers and others. The people liked him very much for his virtues as man and democratic values as a leader. He was their sole leader for their welfare.

20

Bharath was well-trained as a student and well-cultured as man. He took his education in a standard university of values. It was nothing but Shantiniketan where his maternal grandfather, Vishesh learnt the values of distinction. He was the top of the university.

Apart from education, from his father Bharath as man learnt many values and virtues. He shaped himself into a worthy son in the tutelage of his worthy father. Behind his greatness, there was his father, Prashanth.

Bharath excelled his father Prashanth who excelled his father Vishwanth. Bharath made his mother Shasima's dreams come true. For three generations, the legacy of service to farmers welfare continued.

Prashanth and Shasima were happy, as their son, Bharath learnt the values of humanity. They were models and wished their son to be model. Shasima as mother wished that he should be a leader par excellence. She was very happy for he excelled his father Prashanth with his praise-worthy service in the welfare of farmers and others.

Bharath became marriageble. Prashanth and Shasima selected a suitable match named Divya for him. She was educated so well as Bharath.

Bharath's marriage was an unrivalled celebration. Many great people attended and blessed the newly married couple. The guests were happy to see the couple born for each other and made for each other. They passed compliments on their union in marriage,

"We've had a big question, in Bhrath and Divya who excel the other?"

"Yes, we can't say who excel the other."

"They're complementary to each other in all respects."

"They're love birds to set a model to all others."

Bharath was in smiles and had an intimate conversation with his wife Divya, telling her all his plans in life,

"Divya, I want to follow the footsteps of my father, Prashanth and grandfather Vishwanath."

"The most welcome...," said Divya.

"I'm for the welfare of the human race."

"The most welcome...," said Divya.

"I'm for the farmer...belonging to the farmer fraternity. The farmer should make his nation flourish well and it is my wish."

"The most welcome...," said Divya.

"I wish that you should follow my footsteps..."

"Sure, I'll be with you, in your plan and action as Sita is with Sri Rama," said Divya, bowing her head.

"Very glad, Divya..."

Divya cooperated with her husband, Bharath in life. He went ahead in the welfare of humanity. She was to pat him, extending her full support.

When the years passed, his father Prashanth, wanted to retire from his services to the farming community and active participation in politics, the people encouraged Bharath to enter the fields as the people loved him for his extraordiary good qualities, saying,

"Bharath, you're our leader now. In view of your father's retirement from farmers' services and politics, there was the wishful demand of the people not just from our region but also from from Kanyakumari and Kashmir," said the people as they heard the spirited voice and heartthrobs of the people in the nation.

"When my dear people want me to come to politics after my father, I'm ready to enter the field," said Bharath.

"You've the qualities of head and heart as a leader. Your services for the welfare of the people," said the people hopefully.

"Keeping in view the state of politics now, I feel that I'm wanted to join politics. The hearts of the people want to see me as their leader," said Bharath.

"Yes...a true leader to shine after your father is necessary," said the people.

Bharath joined and contested the election and won the election. As parliamentarian and as a minister later, he played an active role in the welfare of the people in the nation.

In the next election, Bharath won the election unanimously. As all liked his leadership, the contestants withdrew from the fray, supporting his leadership morally. He won the election unopposed and set a model to the nation.

Bharath went ahead with his ideals like his father. The eradication of poverty, all citizens as partners in agriculture, literacy, medical facility for all people, free electricity, development of industries and factories and so on were his main ideals to run the government. It was his manifeso but not his show to misguide the people. He went to the people irrespective of their sociological or financial status and became close to all the people from Kanyakumari to Kashmir.

Bharath was the friend of the people, especially the dear farmers. He was for the welfare of the people, especially the poor and the helpless. He was known for justice but not for prejudice.

All the parliamentarians elected Bharath their leader. He was sworn in as the Prime Minister of India amid the big rounds of applause.

Bharath simply said in his maiden speech as the Prime Minister of India,

"I'm for the people...for their welfare...They'll live peacefully, comfortably, and happily in my governance...I'm 'leader' to mean my people, the members of my family."

"I'm a farmer...I'm a farmer's son...I'm a farmer's grandson. I'm for the farmer... for the cause of the farmer. When the farmer is happy in a nation, the nation feels pride and dignity. The famer should be happy in a nation like India. The farmers should never shed tears. They shouldn't commit suicides. All should be happy. All should flourish in our nation."

"Since our nation is an agricultural country, hereafter, we all are farmers having equal share in the land available in India. All must farm the land. Work is worship. Work in the sign of farming is real worship in a nation like India."

"As a farmer, Bharath learned that Vishesh, 'the farmer the great' was my maternal grandfather. I studied about my maternal grand father, Vishesh and his attitude towards his workers. He made all his workers the partners of agriculture. He's my role model," said Bharath.

As the leader of the nation, Bharath studied the reforms of successful leaders across the world and implemented them in his governance.

Success was the synonym of Bharath for he ruled the nation establishing law and order, unity and harmony, joy and peace and so on. He expressed his wishes against all untoward incidents,

"Violence has no room in my peaceful nation. Killings, kidnaps, and so on of worst violence result in bloodshed and tears. Humiliation, discrimination, and marginalization are signs of insult to our nation. Deception, curruption and exploitation are not the words heard in the nation to be model to other nations."

"Woman is on par man in all respects. She should be happy enjoying freedom. Her race shouldn't become victims to rapes...nor to gang-rapes... not to any humiliation."

Bharath wanted to encourage virtues and values, merits, and talents of his people across the nation. He accorded due honour to them by felicitating them on rendering invaluable service in their respective fields who are the real leaders of a nation. To encourage all with their superlative qualities and unrivalled services, Bharath felicitated the best leader, the best farmer, the best scientist, the best doctor, the best teacher, the writer, the best officer, the best social worker, the best actor, the best player and so on in the nation to encourage all to have healthy competition in all the fields in the welfare of the nation.

To encourage the deserving in all fields, Bharath felicitated the people of extraordinary skills and services across the nation in the presence of all dignitaries of the nation. All Indians appreciated his decision.

Bharath felicitated many people of rare distinction like Vishesh, Malli, Dakshinya Deekshith and Harsha Deeksith across the nation for their invaluable services.

Dakshinya Deekshith and Harsha Deekshith as teachers to maintain orphanages in the welfare of abandoned babies. They brought up and imparted valuable education to turn them independent and self-sufficient.

Bharath had special reverence for teachers like Dakshinya Deekshith and Harsha Deekshith for they shatter the ignorance of the people by the light of their wisdom-oriented knowledge.

Bharath felicitated Vishesh, Raithu Bandhu, the leader of farmers, his maternal grandfather for his model service in making the workers as the partners to work in the land of his agricuture. He also set model in making his workers his landowners. He dedicated the rare award to his father, Bharath's great grandfather, Vinshwanth. He felicitated Malli for her sincerity in facing her problems and her service to the welfare of women for enriching the image of woman by not getting ready for foeticide. She therefore set a model to enrich the Indian culture in all respects. She also made her workers partners in her agriculture.

In other fields, Bharath felicitated heartily for their invaluable services in their respective fields. Suryasen and Indrasen heard this news to be ashamed of their evil nature.

The recipients of awards responded in their own ways applauding Bharath for his unrivalled governance.

Vishesh dedicated his award to his father Parameshwar who was his role model as a famer. He felt pride and honor for his ideals.

Malli dedicated the award to Vishesh, her parents and the people of Indrapur. She dedicated her award to Vishesh for he was the person to inspire her to sharing the land to the small farmers and for taking up steps in the welfare of women. She felt that he was worthy of the honor.

In the position of the Hon'ble Prime Minister, Bharath proved his humility and dignity as the sign of his love for humanity.

Bharath touched their feet and sought their blessings. All elderly people on and off the dais showered their blessings on him.

The people showered praises on their beloved leader, Bharath as he enriched and excelled the position of the Prime Minister.

"Long live Bharath... Long live Bharath. You're our leader. Every old, every middle age and every child love you from their hearts. The babies in the lap love to lisp your name in tender smiles. The babies in the wombs love to listen to the progress of Bharath."

All the people across the nation had reverence for Bharath as the leader of the nation to rule excellently well. They were happy for his selfless rule. They praised his rule as Rama Rajya as their nation knew progress, peace, and harmony to mark a model to all other nations.

INTRODUCTION

Dr Katta Rajamouly, M. A., M. Phil., Ph. D., Professor of English by profession is a poet, short story writer, essayist, critic, and translator by passion. He has to his credit sixty books and 344 articles, stories, and poems, widely published in national and international journals and anthologies especially ***The Dance of the Peacock***, an Anthology of English Poetry from India; *Diamond India,* ***Oh My Sweetest Love: A Timeless Treasure, Diamond India: Unheard Songs of Obscure Heroes*** and ***Leisure Pleasure and Contemplation***, and so on of high repute. The well-known historian, PCK Prem's **History of Contemporary INDIAN ENGLISH POETRY: An Appraisal (Part-2)** reflects Dr Rajamouly Katta's extraordinary contributions to English Literature as a poet, short story writer, novelist, and skit writer, dealing with kaleidoscopic themes. He has so far composed 3456 poems (including 1512 Haiku), 198 short stories, nine novels and eighteen skits. Besides these, he has books of criticism, ELT, and communication skills.

Dr Rajamouly is currently working as Professor and Head, Department of English (S & H Dept.) at Ganapathy Engineering College, Vaagdevi Group of Institutions, Warangal, Telangana, India. He has had rich experience in the field of teaching English Language and English Literature at undergraduate and post-graduate levels for more than four decades. During his service from 2005 to 2008, he headed the Department of English at Kakatiya Government UG & PG College, affiliated to Kakatiya University, Warangal. He handled M.A. English literature classes during the tenure at the College. At the same time, he has been teaching the post-graduate students American Literature and Methods of English Teaching in India for more than twenty years at the Warangal Regional Study Centre of Dr B.R. Ambedkar Open University, Hyderabad, Telangana. He has participated in several international and national seminars and workshops. He has delivered several extension and guest lectures at colleges for students and teachers.

As **a son of the Indian soil**, Rajamouly was born in a poverty-stricken family at Ramakrishnapur, a hamlet of Veenavanka Mandal Quarters in Karimnagar district in Telangana state. He grew up to the expectations of his parents, Sri Katta Lashmiveeraiah and Smt. Lakshmikantha. He settled at Excise Colony, Subedari in Hanamkonda, Hanumakonda Dist, (Erstwhile Warangal Urban dist), Telangana State, India.

As **a student**, Rajamouly won the encomiums of his teachers as he proved his academic excellence by standing first in the class every year. He won the prestigious Merit Scholarship during his school and college studies. He has had perseverance and strived hard to pursue his studies barring the hazards and hardships faced by his poor family. He obtained his postgraduate degree, M. A. from Osmania University, Hyderabad; and M. Phil, and Ph.D. from Kakatiya University, Warangal. To his credit, he as a researcher completed research for Ph. D. within two years. His several post-doctoral research articles featured in anthologies and journals of high repute to promote his image as a poet and researcher. He simultaneously pursues his research on poetry as a genre, the art of speaking as the skill of communication skills, and the inconstancies and inadequacies in traditional grammars.

As **a researcher**, he worked on the Movement Poetry for his **M. Phil.** and on the concept of time in the poetry of Philip Larkin for his **Ph. D**. from Kakatiya University, Warangal. He published his research book, ***The Poetry of Philip Larkin: a Critical Study*** (Prestige Publishers, Delhi, 2007), his research on Communication skills, ***The Art of Speaking*** (APH Pub. Corporation, Delhi, 2012). For his post-doctoral research, he deals with ***Inconsistencies and Inadequacies in Traditional Grammars: Critical and Analytical Study***, poetry in ***The Essence of Poetry***, and communication skills in ***The Art of Speaking***. His first research book entitled ***New Renaissance: The Movement*** is for his **M.Phil.** Degree.

As **Professor of English**, Dr Rajamouly produced the course material, ***GRAMMAR*** for undergraduate students (School of Distance and Continuing Education, Kakatiya University, Warangal, 2008) and the books on ELT: ***Learn English through Conversation, The Dictionary of Grammatical and Literary Terms*** (APH Publishing Corporation, New Delhi, 2012) apart from research books. The article on ELT, 'The Pedagogical Approaches from the Learner's Point of View' appeared in the anthology of articles, ***English Language Teaching: Perspectives and Challenges*** and ***The Journal of English Language Teaching*** (V-50-1). His ELT (English Language Teaching) books, ***The English Language Laboratory, Communication Skills par Excellence*** *(Speak and Write Correct and Appropriate English: Encyclopaedia of English Usage)*, ***English at Our Doorstep, Structures of English*** and ***Copious Examples to Learn English*** are to his credit.

As **a critic**, Rajamouly Katta has seventy-five critical articles collected in two collections of critical and general articles, entitled ***My Observations*** and **My Perceptions.** They featured in in anthologies and journals of high repute.

As **a creative writer** of all genres, Rajamouly is interested in writing poetry, short stories, fiction, skits and so on, criticising evils in society. He firmly believes in the fact that a writer is one who holds a mirror in his writings to the society he lives in. He conveys a message to the people of his generation and those of coming generations.

As **a poet**, Dr Rajamouly Katta feels that poetry is for pleasure in reading to lead its readers to a treasure trove of wisdom. In his poetry, he deals with a wide variety of themes underlying life: time, life, love, nature, art, sculpture, the society today, etc., with kaleidoscopic details. He deals with the concepts in the genre of his poetry named **Koels of Paradise**. He has so far written more than **3456** poems collected in eighteen compendiums. He has won the **THIRD** place **five** times in the **POETRY CONTEST** in India (B-category) conducted by **Metverse Muse**. He has **eighteen** poetry compendiums with nine sections. His first juvenile compendium: ***Beauty in Variety*** (Authors Press, 2016), ***Cherished Cherries*** (Authors Press, 2016), ***Petals of Insight*** (Authors Press, 2016), ***Voices of Silence, Kavitanjali: Poem Offerings, Melodies of Hues and Perfumes, New Horizons, Strides of Tides, Moonbeams in Clouds, Beams of Blooms, Lutes of Mute Flutes, Musings Lightnings, Traverse in Time, My Motherland, Symphony in Harmony, Flights to Heights,*** **Musings of Muse** and eighteenth compendium, ***Coo...Coos of Cuckoos***: a collection of 1512 Haiku. in his select poems entitled ***Dazzles of My Musings, Parts1, 2. 3 & 4*** are from eighteen mega anthologies. Reviews on his collections of poems appeared widely in esteemed journals like ***Poetcrit*** and ***Eternity***.

Rajamouly has so far contributed poems, short stories and articles to various **ANTHOLOGIES and journals**. His four poems: 'Gifts of Nature', 'Thrall', 'Twilights' and 'Hell of Earth or Earth of Hell' featured in ***The Dance of the Peacock:*** An Anthology of English Poetry from India, ed. Dr Vivekanand Jha (Canada, Hidden Brook Press: 2013). His poem, 'O, My Love' appeared in the anthology entitled, ***Oh My Sweetest Love***: A Timeless Treasure (Indotech Publications PVT

Ltd, Lathur, 2017), 'Love for the Soil', 'What a Nation!' in ***Diamond India, Unheard Songs of Obscure Heroes***. 'Time,' published in several journals, is anthologized in ***Pleasure, Leisure, and Contemplation*** (Iterative International Publishers, 2022) The critical appreciation of my poem, 'Time' was written by the famous poet, DC Chambial. It appeared in **Contemporary Vibes.** His poem 'The Dance of the Peacock' on ***The Dance of the Peacock*** as an e-review poem attracts the attention of the readers of the anthology online. He also contributed his poems to various **JOURNALS**.

Kakatiya Journal of English Studies (KJES) featured his poems 'Time', 'Democracy Today' and 'Dream and Reality'. ***TRIVENI,*** *a quarterly* featured 'Time', 'On the Move', 'The Past', 'Democracy Today' (Second time), 'Thrall', 'An Imprint of treasures', 'Youths Today', 'My Heart Leaps', 'Ephemeral Glow', 'Imprint of Treasures', 'Source for Inspiration', 'The Ephemeral Glow', 'Triveni' and 'Farewell to Jawans'. **Muse India** also featured the poem, 'Farewell to Soldiers'. ***Commonwealth Review*** from New Delhi published 'What Rules Man's Mind Today', 'True Paradise', 'Fancies' and 'Heavenly Bliss'. ***Muse India,*** e-journal from Hyderabad, featured his poems like 'Earth and Offspring', 'Oracle' and 'Silent' (November-December 2012), 'Lovely Insights', 'Sans You I Am Unknown', 'Tearful Farewell to Jawans', The Flower's Plea', 'The Main Missing', 'The River of Life', To Offer, Not to Hide,' and 'Time's Surprises' (May-June 2019). ***Contemporary Literary Review India*** *(CLRI)* published 'Soil' (Vol,2No.2, Apr=June 2015), 'Memory'& 'Three-in-One' (May 2020), 'Drop pf Water' and Flower's Message ((9-2, May,22). ***Poetcrit*** featured his poems, 'The Farmer' (Jan-Jun 2018, Vol XXXI, No.1), 'Journey of Man' (Jul-Dec 2018, Vol XXXI, No.2), 'Lovely Insights' and 'Sans You, I Am Unknown' (Jul-Dec, 2019, Vol.32, No-1), 'Poet's Fountain' (33-1, Jan-Jun, 2020), 'Most Welcome Guest' and 'See My Smiles'(Jul-Dec,2020, Vol.33 No.2), ***Poetcrit*** (35-1, Jan-Jun Issue) featured 'Time Tree', 'Tears of Flowers', 'Sheens of Greens'. 'Sculptor the Great' and 'Look at It at Its Depth' appeared in **Poetcrit** (Jul22), 'Love Is Emotion', in **Poet-**

crit, (36-1), 'Triple-Facets of Time', 'Seed's Autobiography' in Poetcrit (36-2), 'I swing in Joy', 'Seed's Autobiography' **Poetcrit** (37-1); ***Metverse Muse*** of 51st to 53rd Triple Issue, August 2018 featured six poems: 'Instinct and insight', 'How Excellence is Transience', 'Competition', 'Marriage Flawless', 'Amazing Is Life' and 'Sweet Dream in Life'; and fifteen Haiku pieces: 'Silent Nightingale', 'Knowledge', 'Devotion', 'Character', 'By Actions', 'Man', 'Truth', 'Pure Love', 'Fame', 'Beauty', 'Appearance', 'The Cuckoo', 'Hidden Treasure', 'Ink' and 'Sole Goal'.. ***Metverse Muse*** of 54th to 56th Muse, 2019 featured sixteen poems, 'Nature is Great', 'Life Mirror', 'Sensuous Beauty', 'Teacher', 'Love', 'Friendship', 'Books', 'Education', 'Mind', 'India', 'Poetry', 'They Are My Fellow Beings' (Revised as 'Vicious Circles'), 'Lightning', 'Virtues Vs Vices', 'Tunes of the Heart', 'They Aim Higher', 'Matchless Treasure' and 'Life, Time's Thrall'. ***Metverse Muse*** of 57th to 59th Triple Issue, February 2020 featured his poems: 'Soul', 'Heart', 'Leader', 'Individual', 'Society Today', 'Nation', 'Governance', 'Blood', 'Real Worship', 'Reality', 'Echo from far Realms', 'Poetic Hues', 'Story Message', 'All that Unseen', 'The Same', Goods Sold', 'Good Will', 'Real Judges', 'Ideal Governance', 'Democracy for All', 'Hearts Bloom', 'Poetry', 'Human Excellence', 'Muse, the Spring of Bower'. ***Metverse Muse***, Diamond Jubilee 60th Issue featured his poems, 'Demo'no'cracy', 'Multiple Treasures', ' A Ride by My Side', 'Great Is Man', 'Like the Mermaid', 'Mother', 'Father', 'Corona Virus', 'Fair Is Love in Race', 'Mother Is Crying', 'For Goals in Stature', 'What I love Most', 'Human tree', 'Fears for Flowers', 'Votes for Notes', 'The Song for a Nightingale', 'The Race for the Race'. ***Metverse Muse*** 61st to 63rd Issue, Aril, 2021 featured 'Ode on Spring', 'Sweet Dream in Life' 'Marriage Is Flawless', 'Instinct and Insight', 'Artful Arts', 'Dilemma', Creative Imagination', 'Musings', 'Buddha', 'Wild Growth of Weeds'. ***Metverse Muse*** 64th to 66th Issue September, 2021 published 'Humanity', 'Relation Unique', Words' Value', Spring, Youth Transitory', 'Three Priorities', 'I Lull You'(Lullaby), 'Time Wait' (Lyric), My Heart Feels All', 'I Love Rivers', 'Sculptures for Raptures' (Ode), 'Bond of Love'(Terza Rima), Srirama's Legacies', My Favourite Journal (Ottava Rima),

'Sonnet on Yoga', Sonnet on Love', 'Courts', 'Acts for Peace', 'Mighty Pen', 'Mirror , Teacher', My Dear Cow Stolen' (Elegy), 'Dazzles of Drizzles', 'All to Excel Actual'. ***Metverse Muse*** (67 to 69 Issues, Jan 22) 'Son Indeed', 'Complexities of Relations', 'Abduct of Beauty', 'Not Small, But Big', 'Book Reading', 'Villain', 'Snakes in the Grass', 'Virtues Shine', Dogs' Tails', 'Essence of Poetry' and 'Clarion Call for Peace'. ***Metverse Muse,*** (70-72 issues June 2022), 'Deep Love', 'Image', 'Robust Youth', 'The Irony of Life', 'War Versus Peace', ***Metvesrse Muse*** (73&74 Issues) 'Towards Peaceful Existence', 'Wedding Album', 'Lightnings Musings', 'War Versus Peace' 'True Lovers for Reunion'. **Metverse Muse (**T5th Issue) 'Life Race in Time's Embrace'(Davidian), 'Wars Versus Scars', 'Flower's Life Wish', 'Dreams in Streams', 'Wealth of Treasures', 'Mom! In Your Lap Once Let Me Sleep'. 'To Flower Great', Strike in wakefulness' (Kyriel), 'House in Heaven' (Tanka). 'I Love Poetry'(Limerick), 'Nature Is a Tresure', 'Life through Time', 'A Sonnet on Youth', 'A Moving Sculpture', 'Noblest, the Teacher's Profession' in **Metverse Muse (**T6th-79th Issues), 'My Dear Firefly', 'Milkman's Replies', 'Extra Super Moon', 'Hive of Honeyed Memories', 'Fancy Order Show', 'Musings like Lightning', in **Metverse Muse (**80th-83rd Issue), 'Democracy Today', 'Nature in Stature',Metverse MUSE TOI Amuse', 'Love in Flow to Glow', 'Self-Sacrifice', 'Time', 'To Flowers's True Hues' in Metverse Muse (84-87 Issue); ***Eternity***, A Spiritually Enlightened International Magazine-21, Jan-Apr 2021 Vol. 8, No. 2 featured 'Flower's Message' ***Eternity** (Magazine june22) featured 'From Darkness to Light'.* '***Creation and Criticism*** Vol.6 Joint Issue featured 'Feel Pride', 'Time's Traverse', 'Transient but Excellent', 'Gandhiji's Message', 'Stride of Tides'. **Verbal Art** featured the poem entitled 'All for Oneness'.

As a **short story writer**, Dr Rajamouly has so far written **198** stories on diverse topics to hold mirror to the society today. He named the genre of short stories, **Woodland Streams**. His first collection of eighteen short stories entitled ***Post-Gandhian Days*** (2016) depicts a rich variety of themes related to social evils in the post-independence era. He deals in the stories with communalism, exploitation, corruption, dowry system, indifferent attitude, feticide, falsehood, untouchability, banking system, deception, ragging, false prestige, selfishness, male atrocities, the falling of standards and various other themes in other stories. In the second collection of his eighteen stories, ***Thorns in the Path*** (2017) he deals with the social values of marriage system, gender equality, commitment towards profession, social service, neighbourhood, care for parents, mother's love, education system, honesty against hypocrisy, property rights, welcome changes, administrative lapses, equality, the love of birds towards man and care and concern for parents. His third collection of stories: ***Shadows of Realities,*** deals with social diverse themes. His fourth collection of short stories, ***Naked Truths,*** deals with multiple themes. His fifth collection of short stories, ***Tales for a New Sunrise,*** sheds light on social realities. His sixth collection of eighteen short stories entitled ***Time Tells Tales*** Deals with various aspects. His seventh collection of short stories, entitled ***Marriages Mirages*** deals with marriage and love in the stories... His eighth collection of fifty-four mini stories for children, entitled ***Mini Tales for Tiny Tots,*** dealing with various morals. The sun, the moon, the stars, clouds, rains, animals, birds, trees, reptiles, insects, leaves, plants, weeds, flowers, fruits, seeds, mountains, hills, rocks, wood, dolls, sculptures, chariots and so on are the characters in the mini stories like those in ***Panchatantra***. They speak like human characters. Through the mini stories of one and a half pages each as a lesson to human beings, the evil nature and the negative traits of animals are in dealing in revelation in comparison with those of human beings today. His ninth collection of ***Dazzles of Darkness*** deals with social problems. He contributed several stories in the journals of high repute His first short story, 'The Sons of the 'Soil' featured in ***Muse India,*** e-journal, Hyderabad. The story entitled 'The Writer' appeared in ***Satavahana Journal of English Studies*** (SJES), 'Cute Servant Maid' (Aug 2020) and 'Love, the Key to Life (November 2020), 'Vinamra' (Vol.9, No.4) featured in ***Contemporary Literary***

Review India (CLRI). The journal featured 'The Cute Servant-maid', 'The Victim', 'English', 'Love is the Key to Life' featured in ***Creation and Criticism*** Joint Issue 18&19 October 2020. **Englit** features "Happy Reunion" **in** Jan 2021. "Ransom" in 2022, 'Female Issue' in 2023. His story entitled 'The Scientist' appeared in famous journals like **Phenomenal Literature.**

As **a novelist**, Dr Rajamouly has written nine novels: ***Rajarishi in Quest of Peace, Tears and Smiles of Telangana: Man-for-Man, Smart Child, Tears of Tribal Hamlets, Ebb and Flow*** and ***Shadows Shattered, Life Tides, Dowry Venom,*** and ***Love… Love… Love,*** *depicting diverse themes*. He names the genre **Novel Marvel to Excel.** He deals with social themes in his novels, longer narratives, ***Rajarishi in Quest of Peace*** (Authors Press, New Delhi, 2017) is the first novel to deal with world peace. ***Tears and Smiles of Telangana: Man-for-Man*** *deals with the backdrop with the movement for sperate statehood for Telangana.* His third novel, ***Smart Child,*** deals with the psychology of a child. The fourth novel, ***Tears of Tribal Hamlets,*** portrays the troubles faced by the tribal people as aboriginals. The fifth novel, ***Ebb and Flow*** deals with the movement of women or the safeguarding of their rights in their welfare. His sixth novel, ***Shadows Shattered,*** deals with flaws in the present education system. His seventh novel, ***Life Tides***, deals with the problems of farmers today. His eighth novel ***Dowry Venom*** deals with problem of dowry system. His ninth novel, ***Love… Love… Love*** deals with love marriages. ***Smart Child, Tears of Tribal Hamlet, Ebb and Flow*** and ***Shadows Shattered*** (Yking Books, Jaipur, 2019) His novels focus on man-to-man relations. He is in quest of harmony and peace not just in India but also in the world. Reviews on his novels like ***Rajarishi in Quest of Peace*** featured in ***Poetcrit,*** Jan,'19.

As **a research supervisor**, Dr Rajamouly has been guiding the research scholars of different universities. Under his supervision, nine scholars have so far pursued research on various poets like Tagore and Sarojini Naidu and writers in the areas of short fiction, fiction, ELT, communication skills and so on.

As a **translator**, Dr Rajamouly has so far translated 108 poems of British, American, Indian, and African, Commonwealth literatures into Telugu entitled ***Vaideshiki Kavitha***. He is translating the Telugu philosophical pieces of Veerabrahmendra Swamy's ***KALAGNANAM*** and the poems of Dr C. Narayana Reddy, Prof. Gopi and others into English in his book entitled ***Rays from the Roots***. Dr Rajamouly Katta's English renderings of Dr Narayan Reddy's Telugu poems: 'Absorption', 'Towards Success', 'Fleeting Nature' 'Unfailing Progress', 'Voice Knows No Age' and Mother's Call' and 'Punctuality' and that of Gopi's 'I Do Not let Time Sleep', 'The Flow of Time' of Gopi) featured in the UGC Approved journal, ***Poetcrit*** Jul – Dec18, Jan-Jun20, Jan-Jun,2022, Jul-Dec23, and Jan-Jun24 respectively. His Telugu renderings like 'Sayam Samayam Vanam Chentha Aagadam' the translation of Translation of Robert Frost's poem, 'Stopping by Woods on a Snowy Evening', 'Namrataga Matladu' Translation of the English poem 'Speak Gently' and others appeared in ***Musi***, a popular Telugu monthly from Hyderabad.

As a bilingual writer Rajamouly writes essays, travelogues, poems, one-act plays and so on in his mother tongue, Telugu, the favourite language for his literary pursuits. They have been collected in **Vachana Rachana** (Verse in Free Verse).

As **a skit writer**, Rajamouly Katta has endeavoured to write eighteen skits with diverse themes under the titles, ***Eye Openers, Enlighteners,*** and ***Illuminators***.

As **a critic**, Rajamouly published ***The Poetry of Philip Larkin: A Critical Study*** (Prestige Publishers: New Delhi, 2007). His critical books include his first research entitled ***The New Renaissance: Movement***: Thematic study. His **seventy-five** critical, ELT and social articles appeared ***My Observations*** (Authors Press: New Delhi, 2017), **My Perceptions** and critical journals. He has contributed several articles to various anthologies of critical essays. 'Poetry for Reform' in the poetry of Susheel Kumar Sharma,' appeared in ***Yking Concise Ency-***

clopaedia of Language, Literature and Culture (Yking Books Jaipur., 2014) and 'Quest for Human Values in the Triple Streams of I.V. Chalapathi Rao: An overview' was published in ***Flowering of Indian English Literature: New Dimensions*** (Prestige Books International, New Delhi, 2014). 'Quest of Salvation and liberation in Potana *Bhagavatham*: a series of Interludes' featured in ***Bhakti Movement and Literature: Reforming a Tradition*** (Rawat Publications: New Delhi, 2016). His article, 'The Essence of Poetry', featured in ***Contemporary Vibes*** and TRIVENI. ***My Observations*** (Authors Press, Delhi, 2017) includes all these articles. His article, 'TV Reddy's poetry: A Journey in Quest of Reform with a special reference to Thousand Haiku Pearls', appeared in ***The Poetic Craft of T.V. Reddy*** (Authors Press: New Delhi, 2017). His article, 'Nature: Fairest Eve in Eden' on *TV Reddy's Thousand Haiku Pearls*, featured in ***The Poetry of T.V. Reddy: Critical Study of Human Concerns*** (Modern History Press, Michigan-USA, 2018). 'Ethical Values in Baddhena's Poetry' appeared in ***Satavahana Journal of English Studies*** (1/2019). His critical article, '*Songs of Sonority and Hope*: A memento of past memories' on DC Chambial's poetry appeared in ***Contemporary Literary Review India*** (CLRI), 2020. 'Interpretation of Jen Walls' poems: A Love Poet par Excellence' featured in ***Contemporary Vibes***, Jan-Mar 2021 and in ***Contemporary Review India (CLRI)***, February 2021. 'Manas Bakshi's Poetic Concerns for Human Relations': overview with reference to ***Parnassus Revival*** is under publication in the next issue of ***Poetcrit***. The 'Poetic Art of Birbhadra', with Special Reference to Thematic Variety in ***Prestine Stroke***, featured in the international journal, ***RUMINATIONS,*** Supplementary Edition, New Delhi (December 2018). 'Nature in three Poets of Three Nations: Wordsworth, Cinare and Robert Frost' July-Dec 2018 (XXXI No 2) His research article entitled "Time in TV Reddy's Poetry with reference to Thousand Haiku Pearls in Comparison with Larkin's poetry in ***Poetcrt*** Jul-Dec 2019 Vol:32, No.2. 'Quest for Human Values in Triple Streams of Professor IV Chalapathi Raao: An Overview featured in ***Metverse Muse*** Triple Issue, June 2019. The research articles, 'The Concept of Time in Larkin: An Indian Perspective' featured in ***Poetcrit*** Jan-Jun 2020 Vol.33 No.1.) and 'Bakshi's Poetic Concerns for Human Relations: An Overview with a Special reference to Parnassus of Revival' in ***Poeticrit*** Vol.33-2). 'The Concept of the Past: "The Forgotten Boredom" in the Poetry of Philip Larkin' **Poetcrit** (V-35-No 1) 'The Essence of Poetry' appeared in ***Contemporary Vibes*** in April 2020. 'Pedagogical Approaches from Learner's Point of View' in **EDUTRACKS** (Vol-20, No:7, Oct 2020), 'Speaking: Skill of Skills' appeared in ***EDUTRACKS***: A monthly Scanner (Vol.i9 No 7) and in **Contemporary Literary Review India** (CLRI) (Nov 2020, Vol.7 No, 4). 'Languages Levels', **EDUTRACKS,** Vol:20, No. 6, 'Inconsistent Curriculum all levels' in **EDURACKS,** Vol. 21-no-2, Oct2021, 'Online Education: Smart Alternative to Face-to-face Teaching and Learning in Covid-19 Pandemic Times' in **EDUTRACKS** (Mar 2022), 'Soft Skills for Teachers' **EDUTRACKS**, Vol. 21 No.11 July 2022, 'Is Education Serving its Purpose?' **EDUTRACKS** No. 22 No. 2 Oct 2022, 'Teaching Doctors, and Nurses', **EDUTRACKS**, Vol.22-No.5 Feb2023, 'New Educational Policy in the Age of Science and Technology: Pragmatic and Scientific Approach' **EDUTRACKS** Vol. 23 No.2, Oct2023. 'Language Levels' appeared in ***EDUTRACKS*** (February 2021). Online Education: "Smart" Alternative to Face-to-face Teaching and Leaning in Covid-19 Pandemic Times' in **EDUTRACKS** (March-2022). 'Soft Skills for Teachers'(July,22), Is Education Serving Its Purpose?'(Oct 22) in **EDUTRACKS**. 'New Educational Policy in the Age of Science and Technology: Pragmatic and Scientific Approach' (Oct 2023) His Critical appreciation of Robert Best's poem, 'A Perfect Start to Autumn' appeared in **The Shamanic Poet** (2nd Sep 2021) in UK.

***POEMS. ARTICLES, REVIEWS & APPRECIATIONS IN JOURNALS AND ANTHOLOGIES**

'Time' appeared in Journals and Anthologies

1. The Poem, 'TIME' featured in **Kakatiya Journal of English Studies,** Vol.19, 1999

2. TRIVENI, Literary and Cultural Journal, Oct-Dec 2005

3. Anthology: **Poems for Leisure, Pleasure, & Contemplation** (Print Publication, New Delhi, '22)

The poem, 'Democracy Today' appeared in TWO journals:

1. **Kakatiya Journal of English Studies,** Vol.20, 2000
2. **TRIVENI,** Oct-Dec 2009 &**TRIVENI**, Apr-Jun 2011

The poem, **'Farewell to Jawans'** featured in **TWO** journals

1. Muse India, e-journal, Vol. 85, May-June 2019

2. TRIVENI Literary and Cultural Journal, Jan-Mar 2019

Certain Articles Appeared in Various Anthologies and Journals:

'Human Values in Triple Streams of IV Chalapati Rao' in FIVE Anthologies/ journals

Anthologies: **1. Flowering of Indian English Literature,** Authors Press, 2013

1. **My Observations, Authors Press,** New Delhi, 2017.

Journals: **3. Metverse Muse** 54th to 56th Triple Issue June 2019

4. TRIVENI, Jan-Mar 2014

5. TRIVENI, Vol.2, 845 Apr-Jun 2019

'The Essence of Poetry' featured in THREE Anthologies/Journals

1. **My Observations, Authors Press,** New Delhi, 2017

Journals: 2. **Contemporary Vibes,** April-June 2020

1. **TRIVENI,** Literary and Cultural, Jul-Sep.2014

'Time in the Poetry of Philip Larkin: An Indian Perspective' published in THREE Anthologies/Journals

Anthology, 1. **My Observations, Authors Press,** New Delhi, 2017

Journals: 2. **Poetcrit, Jan-Jun2020 pp28-33**

3 **TRIVENI,** Literary and Cultural Journal Apr-Jun,2012'

Potana Bhagavatam: A Series of Interludes in Quest of Salvation featured in THREE

Anthologies: **1. Bakti Movement and Literature: Re-forming,** Rawat Publications, New Delhi pp120-129

2. My Observations, Authors Press, New Delhi, 2017.

Journals: 3. **TRIVENI**, Literary and Cultural, Jul-Sep 2013

'Gitanjali, Bhakti Lyrics' featured in THREE

Anthologies: 1. **My Observations,** Authors Press, New Delhi, 2017.

Journals, **2. Kakatiya Journal of English Studies**, Vol 31. 2012

3.TRIVENI, Vol 82 Oct-Dc, 2012

'The Art of Speaking, the Skill of Skills: A comprehensive Study' featured in THREE

Anthologies: 1. **My Observations,** Authors Press, New Delhi, 2017.

Journals: 2. **EDUTRACKS,** Monthly Scanner, Trends in Edn.Vol.19, No7, 2020

3. **Contemporary Literary Review India** (CLRI)

Article. 'Pedagogical Approached from Learner's Point of View' published in THREE:

Anthologies: 1. **Strengthening of English Language and Learning:** Issues and Challenges, 2014

Journals: 2. **Journal of English Language Teaching**-50, 2012,

3.EDUTRACKS Oct-2020

'Manas Bakskhi's Poetic Concerns for Human Relations' published In TWO

Poecrit, Jul-Dec 2020 Vol-33, No.2

Creation and Criticism, April 2020

'Soft Skills for Doctors', published in THREE books.

My Observations, Authors Press, 2016

Ebb and Flow (Novel) , Yking Books, Jaipur, 2019 as an address of a doctors

'Teaching Doctors and Nurses', **Edutracks,** Vol.22, No 6, February 2013

***REVIEWS of his literary works, Cherished Cherries appeared in TWO**

Poetcrit, Cherished Cherries, a Review of Rajamouly by DC Chambial Vol-32-1

Eternity, A Spiritually Enlightened Journal, Vol-9-1

Prof. D C Chambial's critical interpretation of the poem, 'TIME' of his second anthology ***Cherished Cherries*** featured in ***Contemporary Vibes*** in October 2019.

***Critical Appreciations in Collections: Dr Rajamouly contributions to Anthologies:**

1.The Shamanic Poet UK, LONDON

Dr Rajamouly's critical appreciation of Robert Best's poem, 'A Perfect Start to Autumn', Adminshaman 6th September 2021 in ***The Shamanic Poet, Carving wisdom on to the back of the World* (UK,**

LONDON)

2. Poetry of DC Chambial (Collection of Critical Essays)

'Songs of Sonority and Hope', A Memento of Memories pp176-96

''Life in Communion with Nature": Poetry of DC Chambial, PP188-92

Yking Concise Encyclopaedia of ***Language, Literature and Culture, Jaipur India***

'Poetry for Reform': *The Poetry of Susheel Kumar Sharm, pp54-6*

4.Flowering of Indian English Literature: New Dimensions, Prestige Books International, Delhi, Sidney

'Human Values in Triple Streams of IV Chalapati Rao' pp278-285

5.Bhakti Movement and Literature: Re-forming a Tradition, Rawat Publications, New Delhi,2016

'Quest for Salvation in Potana Bhagavatam *pp120-124*

6.The poetic Art TV Reddy: New Perspectives with special refence to 1000Haik Pearls

'TV Reddy's Poetry: A Journey in Quest of Reform'

7.TV Reddy's Poetry: A Critical Study of Human Concerns, *Modern History Press, AnnAvpor, Michigan USA, 2017 'Nature, the Fairest of the Eve in Eden: TV Reddy's thousand Haku Pearls'*

8.Strengthening of English Language and Learning UGC Excel Publishers, New Delhi

"Pedagogical Approaches from Learner's Point of View" 2012, 50-1 PP 34-41.

9. ***Contemporary Vibes***

'The Essence of Poetry' featured in (Vol. 15, Issue 59) April-June 2020.

Dr Rajamouly Katta's Poem entitled "Time" of ***Cherished Cherries,*** Authors Press, New Delhi, 2016 by "Prof. D.C Chambial, in Oct-Dec 2019,

Interpretation of Jen Walls' poems, 'Love Poet par Excellence', Jan-Mar 2021.

10.*Commonwealth Review, New Delhi*

Poems 'What Rules Man's Mind Today' (XV-2), 'True Paradise' (X-2), 'Fancies' (XXI-2), 'Heavenly Bliss' (XXI-2) in different issues

11. Creation and Criticism

"Manas Bakshi's Poetic Concerns for Human Relations" an Overview, With Special Reference to Parnassus of Revival, **Creation and Criticism,** 22June 2020.

With poems in Vol.6, Joint Issue, Jan-April, 2021: 'Feel Pride', 'Time's Traverse', 'Transient but Excellent', 'Gandhiji's Message', 'Strides oof Tides'

'The Victim' (Story) Vol.5 Joint Issue, 18&19, Oct 2020 'English' (Story) Vol,6 Joint Issue, July-October 2021.

12. *TRIVENI, Cultural, literary Quarterly, Hyderabad*

'The Drama of Man in Nature, Treatment of Nature in Frost and Wordsworth', Vo. 81 No.1 Jan-Mar 2012.

'Time in the Poetry of Philip Larkin', Vol. 81 No. 2 Apr-Jun 2012

'The Quest for Human Values in Triple Streams of IV Chalapathi Rao', Vol. 83 No.1 Jan-Mar 2014

'Quest for salvation in *Potana Bhagavatham*: A Series of Interludes in quest of Salvation or Liberation' Vol. 82 No 3 Jul-Sep 2013

'Quest for Human Values in Triple Streams of **TRIVENI,** An Overview', No 1Vol. 83 No.1 Jan-Mar 2014.'The Essence of Poetry', No 3 Vol. 83 Jul-Sep 2014.

'Themes in the Poetry of Philip Larkin', No-2 Vol. 86 Apr-Jun 2017.

'The Concept of Time in Larkin, an Indian Perspective', Vol. 84 Apr-Jun 2015.

'Yogi Vemana the Poet of Commonman', No-4, Vol. 84 Oct-Dec 2015.

'Narayana Reddy- Ci Na Re: A Poet and Lyricist', No-3 Vol. 86 Jan-Mar 2017. 'Quest for Human Values in the Triple Streams of I.V. Chalapathi Rao' Prof. IV Chalapathi Rao, Commenmorative Edition.Vol.2, No.2, Apr-Jun 2019 (2nd time)

'Nature in Three Poets of Three Nations: Wordsworth, Cinare and Frost' (Apr-Jun'18)

'The Poetic Art of Birbhadra Karkidoli: Thematic variety in Pristine Stroke, (90-1 Jan'22)

With thirteen poems: 'Time', 'On the Move', 'Democracy Today', 'The Past', Democracy Today (Second Time), 'Thrall', 'Youth Today', 'My Heart Leaps', 'Ephemeral Glow', Imprint of Treasures', 'Source for Inspiration', 'TRIVENI', Tearful Farewell to Soldiers' in different issues.

13. Kakatiya Journal of English Studies (KJES)

'*Gitanjali*: Bakti Lyrics of Humanistic and realistic Approaches', 31/2012

'Agnosticism in the poetry of Philip Larkin': A Note on 'Church Going', 32/2013

'Larkin's Poetic Credo and Rhetoric' 36/2017

'Time', Dream and Reality' and 'Democracy Today' in two different issues

14. Satavahana Journal of English studies **(SJES)**

Critical Article, "Ethical Values in ***Sumati Satakam***", Vol 2019

"The story entitled "Writer", Vol 2019

15. Poetcrit (International Journal), Prints Publication PVT Ltd, New Delhi

'Nature in three Poets of Three Nations: Wordsworth, Cinare and Robert Frost', July-Dec 2018 (XXXI No 2)

'Time in the Poetry of T.V. Reddy in comparison with Philip Larkin' published in Jan-June 2019 (XXXII)

The research article entitled 'The Concept of Time in Larkin: An Indian Perspective' featured in ***Poetcrit*** Jan-Jun 2020 Vol.33 No.1.

'The research article 'Bakshi's Poetic Concern for Human Relation: An Overview with a Special reference to Parnassus of Revival' Vol.33-2.

'The Concept of the Past: "The Forgotten Boredom", the Poetry of Philip Larkin', (35-1).

'Time: Double Edged Weapon: An Overview of Larkin's Poetry, Poetcrit, Vol. 36-2 2023

'The Theme of Death', Life in Time: Larkin's Perspective of Death 'A Black-sailed Familiar' Poetcrit, 37-1, Jan-Jun 2024.

With poems: 'Farmer'(XXXI-1), 'The Journey of Man' (XXXI-2), 'Lovely Insights (XXXII-1), 'Sans You, I AM Nothing' (XXXII -1), 'Poet's Fountain' (33-1), Most Welcome Guest' (33-2), See, 'My Smiles' (33-2), 'Time Tree' (35-1), 'Tears of Flowers' (35-1), Sheens of Greens' (35-1), Sculptor the Great' (35-2), 'Look at its Depth' (35-2), 'Love Is Emotion' (36-1), Triple-facets of Time', Seed's Autobiography.' Poetcrit, Vol.36, No.2 2023.

With translations of poems of Dr C. Narayana Reddy from Telugu to English: 'Absorption', 'Towards Success', Fleeting Nature', 'Unfailing

Progress', and 'Voice Knows No Age' And the of Prof. N. Gopi, 'I Do not Let Time Sleep'.

16. ***Muse India, e-journal***

'Cinare and Philp Larkin: Comparative Study' Muse India@75 Sep-Oct 2017

With several poems: 'Earth and Offsprings', 'Oracle', and 'Silent' (Nov-Dec, 2012), 'Lovely Insights', 'Sans I Am Unknown', Tearful Farewell to the Jawans', 'The Flower's Plea', 'The Main Missing', 'The River of Life', 'To Offer Not to Hide', Time's Surprises' (Vol-85, May-jun,2019)

'Sons of the Soil' (Story)Jul-Aug2012.

17. ***Ruminations*** (International Journal) *Supplementary Edition*

The 'Poetic Art of Birbhadra', with Special Reference to Thematic Variety in ***Prestine Stroke,*** featured in the international journal, ***Ruminations,*** Supplementary Edition, New Delhi (December 2018). pp 184-202

18. Metverse Muse (International) Vishakhapatnam

The esteemed Editor Dr H. Tulsi published conferred on Dr Rajamouly Katta honorary life membership, publishing 145 poems so far. His full-length introduction along with seven poems and fifteen haiku poems featured ***Metverse Muse,*** Nos. 51st to 53rd Triple Issue, 2018.

On Prof IV Chalapati Rao, my widely published article 'Human Values in Triple Streams of *TRIVENI' appeared in* ***Metverse Muse,*** *Nos 54th to 56th (June,2019)*

19. Contemporary Literary Review India, Pune, Maharashtra

'Songs of Sonority and Hope, A Memento of Past Memories', A Critical Overview of DC Chambial's Poetry (CLRI) Vol7, No.1, February 2020

The story entitled 'Cute Servant-maid' featured in Contemporary Literary Review, India (CLRI) August 2020.

The story entitled, 'Love Is the Key to Life', Vol.7, No.4 November 2020

The article, 'Speaking, the Skill of Skills', A comprehensive Study, Vol.7, No.4, Nov'20

The Critical Appreciation, 'The Poems of Jen Walls, An Interpretation (8-1), Feb.2021

The story, entitled 'Vinamra' Vol.9, No. 4.

With four poems: 'Soil', 'Memory: Three-in-One', 'Drop of Water', Flower's Message in three different journals.

20. *EDUTRACKS*: A Monthly Scanner of Trends in Education, Nilkamal publications

Nine articles: 'Pedagogical Approaches from Learner's Point of View', 'Speaking: Skill of Skills', 'Languages Levels', 'Inconsistent Curriculum all levels', 'Online Education: Smart Alternative to Face-to-face Teaching and Learning in Covid-Pandemic Times', 'Soft Skills for Teachers' 'Is Education Serving its Purpose?', 'Teaching Doctors, and Nurses', 'New Educational Policy in the Age of Science and Technology: Pragmatic and Scientific Approach'.

20. Journal of English Language Teaching, Bangaluru

'Pedagogical Approaches to Learner's Point of Views' 50-1 PP34-41, 2012

21. Phenomenal Literature, Critical Journal devoted to Lang. & Literature (Ed. Vivekanand Jha)

'The Scientist' (Story) Vol.7I, issue-2, March, '23

22. Verbal Art, a Global Journal devoted to poets and Poetry (Ed. Vivekanand Jha)

Poem, 'All for Oneness' (Poem) Vol.6, Issue-2, March, '23

23. **Eternity,** A Spiritually enlightened International Magazine, Bubaneshwar, India

Review of his Poetry Collection, Cherished Cherries

Poems: 'Flower's Message', 'Light in Darkness'

24. **Englit (International Journal) Kolkata**

'Happy Reunion' (Story) 1st issue, 2021, 'Ransom' (Story) 2nd issue, 2022, 'Female Issue' (Story) Third Issue 2023.

***Rajamouly's poems in Various Anthologies**

25. 'Dance of the Peacock: An Anthology of English Poetry from India, Hidden Brook Press, '13

'Gifts of Nature', 'Twilights', 'Thrall', Hell of Earth or Earth of Hell' pp207210

26. O! My Sweetest Love: A timeless Treasure, Indotecj, Publications, 2022 (Poen 'O! My Love')

27. Diamond India-Unheard Songs of Obscure Heroes, IIP International Publishers, 2022

'What a Nation!', 'Love for the Soil' p.73

28. Leisure, Pleasure, and Contemplation Anthology Print Publishers Pvt, Ltd, New Delhi, 2022

'Time'

29. Musi (Telugu Journal)

Poems of English into Telugu appeared in **Musi**, a popular journal from Hyderabad.

30. Social Articles appeared in TRIVENI and anthologies *My Observations* and *My Perceptions*.

As **an essayist** too, Dr Rajamouly Katta as a man of social consciousness has written a considerable number of social articles to his credit. They have been collected in **My Observations** and **My Perceptions**. His essays: 'Diminishing Beauty', 'Isn't Man a Social Animal?', 'The Fancy-dress Show', 'What Makes Man Respectable' 'False Colors Vanish', 'Demons, Humans and Divines', 'Forms of Intolerance', 'Education', 'Marriage', etc., reflect multifarious aspects of the society today. 'Diminishing Beauty', 'Isn't Man a Social Animal?' and so on were published. '*Batukamma*: A Coronation to Flowers' featured in the journals like ***TRIVENI***. Vol.3'19.

rajamoulykatta@gmail.com

WRITINGS OF THE WRITER

Critiques Creative (Poetry, short stories, novel, skits, translations) and **Critical Writings**

***The Creative Craft of Rajamouly Katta*: Redefined Sensibilities and Realities**

Poetry: **Koels of Paradise**

Beauty in Variety (The Juvenile Collection of Nine Anthologies) Authors Press, New Delhi, 2016

Cherished Cherries (The Collection of nine Anthologies) Authors Press, New Delhi, 2016

Petals of Insight (The collection of nine anthologies) Authors Press, New Delhi, 2016

Voices of Silence (The collection of nine anthologies)

Myriad Hues of Varied Melodies (The Collection of nine anthologies)

Kavitanjali: Poem Offerings (The Collection of nine anthologies) Yking Books, Jaipur, 2019

New Horizons (The Collection of Nine Anthologies)

Strides of Tides (The Collection of Nine Anthologies)

Moon Beams in Clouds (The Collection of Nine Anthologies)

Beams of Blooms (The Collection of Nine Anthologies)

Notes of Mute Lutes (The Collection of Nine Anthologies)

Musings, Lightnings (The Collection of Nine Anthologies)

Traverse in Time (The Collection of Nine Anthologies)

My Motherland (The Collection of Nine Anthologies)

Symphony in Harmony (The Collection of Nine Anthologies)

Fights to Heights (The Collection of Nine Anthologies)

Musings of Muse (The Collection of Nine Anthologies)

Coo...Coos of Cuckoos (The Collection of Nine Anthologies)

Dazzles of My Musings: Select Poems Parts 1, 2, 3 &4 (From 18 Anthologies)

Short Stories: **Woodland Streams**

Post Gandhian Days, a Collection of Eighteen Stories Authors Press, New Delhi, 2016

Thorns in the Path, a Collection of Eighteen Stories Authors Press, New Delhi, 2017

Shadows of Realities, a Collection of Eighteen Stories Yking Books, Jaipur 2019

Naked Truths, a Collection of Eighteen Stories Yking Books, Jaipur 2019

Tales for a New Sunrise, a Collection of Eighteen Stories Yking Books, Jaipur 2019

Time Tells Tales, a Collection of Eighteen Stories Yking Books, Jaipur 2019

Marriages, Mirages, a Collection of Eighteen Stories

Mini Tales for Tiny Tots, a Collection of fifty-four stories

Dazzles of Darkness, a Collection of Eighteen Stories

Beauties of Realities: Select Short Stories from six Anthologies

Novel: **Marvel to Excel**

Rajarishi in Quest of Peace Authors Press, New Delhi, 2016

Tears and Smiles of Telangana: Man-far-Man Yking Books, Jaipur 2019

Smart Child Yking Books, Jaipur 2019

Tears of Tribal Hamlets Yking Books, Jaipur 2019

Ebb and Flow Yking Books, Jaipur 2019

Shadows Shattered Yking Books, Jaipur 2019

Life Tides

Dowry Venom

Love... Love... Love

Skits: **Visual Wonders**

Eye-openers: Nine Skits, **Enlighteners:** Nine Skits, **Illuminators:** Nine Skits

Criticism: **Crystal Critiques**

The Poetry of Philip Larkin: a Critical Study Prestige Publishers, New Delhi, 2007

My Observations, a Collection of Critical, ELT,

Social & post-doctoral research articles Authors Press, New Delhi, 2016

***Evolution in the Poetry of Philip Larkin*: A Thematic Study**

My Perceptions, A Collection of Post Doctoral &other Articles

Translations: **Radiant Renderings**

Kavita Vaideshiki (108 English poems into Telugu)

The Rays from the Roots (108 Telugu Poems into English)

Veman's and Baddhena's Poetry (108 Telugu Poems into English)

Post-doctoral Research: **New Sunrises**

The Art of Speaking: The Art of Arts APH Publishing Corporation, New Delhi, 2012

Inconsistencies and Inadequacies in Traditional Grammars

The Essence of Poetry

Methods of Teaching: **Pedagogy Panorama**

The Dictionary of

Grammatical and Literary Terms APH Publishing Corporation, New Delhi, 2012

English Language Laboratory

English at Our Doorstep

Structures of English

Learn English through Copious Examples

Learn English through Language Games

Communication Skills: **Sweet Tongues**

Learn English through Conversation APH Publishing Corporation, New Delhi, 2012

Communication Skills par Excellence

MY SHORT BIO:

Dr Katta Rajamouly, Professor and Head, Department of English (S.H), Ganapathy Engineering College, Warangal has 64 books of all genres to his credit. ***The Poetry of Philip Larkin: a Critical Study, My Observations, My Perceptions, The Movement,*** **Art of Speaking, the Skill of Skills,** ***Learn English through Conversation, Grammar*** and ***Dictionary of Grammatical and Literary Terms*** are his critical and ELT Books. As a poet, he has won the **THIRD** place in the poetry contest from India. He has books of Poetry: ***Beauty in Variety, Cherished Cherries, Petals of Insight, Voices and silences, Myriad Hues of Varied Melodies, Kavitanjali: Poem Offerings, New Horizons, Strides of Tides, Moon Beams in Clouds, Beams of Blooms, Notes of Mute Lutes, Musings, Lightnings, Traverse in Time, My Motherland, Symphony in Harmony, Fights to Heights, Musings of Muse, Coo...Coos of Cuckoos, Dazzles of My Musings***: **Select Poems (1, 2, 3 &4),** his short story collections ***Post Gandhian Days, Thorns in the Path, Shadows of Realities, Naked Truths, Tales for a New Sunrise***; and ***Time Tells Tales, Marriages, Mirages; Mini Tales for Tiny Tots, Beams in Darkness,*** and ***Beauties of Realities,*** and novels ***Rajarishi in Quest of Peace, Tears and Smiles of Telangana,, Smart Child, Tears of Tribal Hamlets, Ebbs and Tides*** and ***Shadows Shattered, Life Tides, Dowry Venom*** and ***Love...***

Love...Love. His 18 skits have been collected in from **Eye-openers, Enlighteners, & Illuminators** His research books ***Essence of Poetry, Inconsistencies and Inadequacies in Traditional grammars*** and his ELT books ***English Language Laboratory, Communication Skills par Excellence, English at Our Doorstep, Structures of English, Grammar in Charts*** and ***Learn English through Copious Examples,*** and translation books: ***Kavita Vaideshiki*** (108 English poems into Telugu) ***The Rays from the Roots*** (108 Telugu Poems into English) ***Vemana's and Baddhena's Poetry*** (108 Telugu Poems into English), His poems appeared in ***The Dance of the Peacock,*** Indian English Anthology (ed. by Vivekanand Jha), and ***O Sweetest Love.*** His review poem 'The Dance of the Peacock' on ***The Dance of the Peacock*** *is* in its review column. He has so far published **344** poems, short stories, articles, and translations in reputed journals: ***The Shamanic Poet*** (UK), ***Contemporary Vibes, Muse India, Poetcrit, Metverese Muse, TRIVENI, Ruminations, Creation and Criticism, Kakatiya Journal of English Studies, EduTracks, Satavahana Journal of English Studies, Eternity, Musi*** and other journals. The reviews of his books appeared in journals like ***Poetcrit.***

rajamoulykatta@gmail.com

Dr. Rajamouly Katta, M.A., M. Phil., Ph. D., Professor of English by profession and poet, short story writer, novelist, writer, critic and translator by predilection, has to his credit **64** books of all genres and **344** poems, short stories, articles and translations published in journals and anthologies of high repute. He has so far written **3456** poems collected in **18** anthologies, **200** short stories in **9** anthologies, **nine** novels **18** skits. ***Creative Craft of Dr. Rajamouly Katta*: Sensibilities and Realities** is a collection of articles on his works. As a poet, he has won **THIRD** Place **FIVE** times in Poetry Contest in India conducted by *Metverse Muse*. rajamoulykatta@gmail.com

Made in the USA
Monee, IL
03 November 2026

67381350R00115